Antipodal Vivaldi Antennas for Microwave Imaging of Construction Materials and Structures

Mahdi Moosazadeh

Antipodal Vivaldi Antennas for Microwave Imaging of Construction Materials and Structures

Springer

Mahdi Moosazadeh
Center for Infrastructure Engineering, School of Computing,
Engineering and Mathematics,
Western Sydney University
Penrith, NSW, Australia

ISBN 978-3-030-05565-3 ISBN 978-3-030-05566-0 (eBook)
https://doi.org/10.1007/978-3-030-05566-0

Library of Congress Control Number: 2018967687

This Springer imprint is published by the registered company Springer Nature Switzerland AG
The registered company address is: Gewerbestrasse 11, 6330 Cham, Switzerland

To Sahebaz Zaman

Preface

Quick development of microwave non-destructive testing and evaluation (NDT&E) of materials and structures has increased the demand on improved microwave measurement methods and techniques. Of particular interest for civil engineering applications is the use of ultra-wideband (UWB) measurement systems that can provide the detection of defects and damages in construction materials and structures using imaging methods and techniques. The development of UWB antennas for these systems with proper dimensions and high performance is one of the major challenges. The success of applying UWB microwave technique is dependent on the operating frequency utilised for specified material under test. The main aim of this research is to develop and apply novel UWB antipodal Vivaldi antennas (AVAs) for a high-resolution detection of defects and damages in composite materials and structures. The following main contributions have been made in this book.

The first contribution is the development of a small and compact AVA with relatively high gain at the highest operating frequency and front-to-back (F-to-B) ratio, low E-field tilt of beam, narrow half-power beam width (HPBW), low cross-polarisation and side lobe levels and wide operating bandwidth (5–50 GHz) for microwave and millimetre wave imaging of relatively low loss materials. Capability of the proposed antenna for the detection and evaluation of targets embedded in construction foam and in a plasterboard sheet layered structure at different depths is demonstrated.

The second contribution is the development of an AVA with trapezoid-shaped dielectric lens to achieve high F-to-B ratio, low E-field tilt of beam, narrow HPBW and wide operating bandwidth (3.4–40 GHz) for microwave and millimetre wave imaging of loss materials. High gain at the lowest operating frequency is achieved by the use of periodic slit edge. Applicability of the proposed antenna for imaging of layered plasterboard sheets with air-filled cavity is presented.

The third contribution is the development of an AVA designed to operate at frequency range from 2 to 27 GHz and high gain at low frequencies for imaging of construction materials and structures. To extend the low end of frequency band down to 2 GHz, inner edges of the top and bottom radiators have been bent. Applicability of the proposed antenna for UWB imaging of construction materials and structures is demonstrated to highlight the capability of the proposed antenna with the imaging system for the detection of flaws such as void inside construction materials with high-range resolution and deep penetration depth.

In the imaging of high loss materials such as concrete-based ones for the purpose of void detection inside concrete, lower frequency, for example, 1 GHz, is desired to provide deep penetration depth. The fourth contribution is the design of an AVA with bending inner edge technique applied on top and bottom radiators to lower the low end of frequency to 1 GHz. To elevate antenna gain at high frequency, elliptical-shaped dielectric lens has been used. The capability of the proposed antenna for UWB microwave imaging of concrete-based specimens for the detection of voids inside concrete is presented.

The fifth contribution is the design of an AVA with comb-shaped slits applied on edges of the radiators to enhance antenna gain at low frequencies. The capability of the proposed antenna for UWB imaging of concrete beams for the purpose of the detection of voids is demonstrated. The results of microwave imaging of voids inside concrete at different standoff distances between the proposed antenna and the surface of concrete are also presented.

Penrith, NSW, Australia Mahdi Moosazadeh

Acknowledgements

In the name of God, the Beneficent the Merciful

I am grateful to my principal supervisor, Associate Professor Sergiy Kharkivskiy (Sergey Kharkovsky), for his supervision, continuous support, encouragement and recommendations (unfortunately, he passed away on 7 September 2017). Special thanks to my co-supervisor Professor Bijan Samali for his support and advices. The present book would not have been possible without their technical insight. I would like to thank all the academic, administrative and technical staff in the Centre for Infrastructure Engineering at Western Sydney University.

Special thanks go to the technical staff, Mr. Ranjith Ratnayake, and IT service, Mr. Nathan McKinlay, for their assistance and technical support in the experimental programme and software support.

I would like to thank my parents, Mr. Sohrab Moosazadeh and Mrs. Robabeh Ahmadpour AliAbadi, for encouraging me and paying out so much that I can focus on my study. Special gratitude and love to my wife, Mrs. Zahra Esmati, for her continuous patience and support and for standing by me and cheering me up through the good and bad times. I also would like to thank my sister, Mrs. Mina Moosazadeh, and my brother, Mr. Mohammad Ali Moosazadeh, for encouraging and helping me. Without the moral and emotional support of my family, this work would not have been possible.

I would also like to express my gratitude to the publisher, Springer, for accepting this book: Brian Halm (*Production Contact*), Michael Luby (*Responsible Editor*), Nicole Lowary (*Assistant Editor*), and Cynthya Pushparaj (*Production Editor*).

List of Publications

Peer-Reviewed Journal Papers

1. M. Moosazadeh and S. Kharkovsky, "Improved radiation characteristics of small UWB antipodal Vivaldi antenna for microwave and millimetre wave imaging applications", *IEEE Antennas and Wireless Propagation Letters*, vol. 16, pp. 1961–1964, 2017.
2. M. Moosazadeh, S. Kharkovsky, J.T. Case, and B. Samali, "UWB antipodal Vivaldi antenna with improved gain for microwave imaging of construction materials and structures", *Microwave and Optical Technology Letters*, Vol. 59, No. 6, pp. 1259–1264, June 2017.
3. M. Moosazadeh, S. Kharkovsky, J.T. Case, and B. Samali, "'Antipodal Vivaldi antenna with improved radiation characteristics for civil engineering applications", *IET Microw. Antennas Propag.*, vol. 11, no. 6, pp. 796–803, 2017.
4. M. Moosazadeh, S. Kharkovsky, J.T. Case, and B. Samali, "Miniaturized UWB antipodal Vivaldi antenna and its application for detection of void inside Concrete Specimens", *IEEE Antennas and Wireless Propagation Letters*, vol. 16, pp. 1317–1320, 2017.
5. M. Moosazadeh and S. Kharkovsky, "A compact high-gain and front-to-back ratio elliptically-tapered antipodal Vivaldi antenna with trapezoid-shaped dielectric lens", *IEEE Antennas and Wireless Propagation Letters*, vol. 15, pp. 552–555, 2016.
6. M. Moosazadeh, S. Kharkovsky, and J. T. Case, "Microwave and millimetre wave antipodal Vivaldi antenna with trapezoid-shaped dielectric lens for imaging of construction materials", *IET Microw. Antennas Propag.*, vol. 10, no. 3, pp. 301–309, 2016.
7. M. Moosazadeh and S. Kharkovsky, "Development of the antipodal Vivaldi antenna for detection of cracks inside concrete members", *Microwave and Optical Technology Letters*, Vol. 57, No. 7, pp. 1573–1578, July 2015.
8. M. Moosazadeh and S. Kharkovsky, "Compact and small planar monopole antenna with symmetrical L-and U-shaped slots for WLAN/WiMAX applications", *IEEE Antennas and Wireless Propagation Letters*, Vol. 13, pp. 388–391, 2014.

Refereed Conference Papers

1. M. Moosazadeh, S. Kharkovsky, Z. Esmati and B. Samali, "UWB elliptically-tapered antipodal Vivaldi antenna for microwave imaging applications", 18th *IEEE International Conference on Antennas Propagation in Wireless Communications (IEEE-APWC),* 2016, pp. 102–105, 19–23 September 2016, Cairns, Australia.
2. M. Moosazadeh and S. Kharkovsky, "Design of ultra-wideband antipodal Vivaldi antenna for microwave imaging applications", 15th *IEEE International Conference on Ubiquitous Wireless Broadband (ICUWB 2015),* 2015, pp. 1–4, 4–7 October 2015, Montreal, Canada.

Contents

Abbreviations

2D	Two-Dimensional
3D	Three-Dimensional
AVA	Antipodal Vivaldi Antenna
BAVA	Balanced Antipodal Vivaldi Antenna
CST	Computer Simulation Technology
CPW	Coplanar Waveguide
EMW	Electromagnetic Wave
FCC	Federal Communications Commission
F-to-B	Front-to-Back Ratio
GPR	Ground-Penetrating Radar
HPBW	Half-Power Beam Width
MMW	Microwave and Millimetre Wave
MUT	Material Under Test
MWS	Microwave Studio
NDT&E	Non-destructive Testing and Evaluation
OBW	Operating Bandwidth
OEW	Open-Ended Waveguide
PCB	Printed Circuit Board
PNA	Performance Network Analyser
SAR	Synthetic Aperture Radar
TWI	Through-Wall Imaging
UWB	Ultra-Wideband

Chapter 1
Introduction

This chapter presents the background, advantages and applications of microwave and millimetre wave imaging for construction materials and structures. The research objectives of this research, book organisation and original contributions of the book are also provided.

1.1 Background, Advantages and Applications of Microwave Imaging

Nowadays, microwave non-destructive testing and evaluation (NDT&E) technology has undergone a revolutionary transformation by providing valuable images that monitor and diagnose the quality assessment of materials and structures. One of the applications of NDT&E is the measurement of the state of reinforcing steel bars in the concrete (Roqueta et al. 2012; Kharkovsky et al. 2012). For this purpose, UWB techniques and methods have been used. They have also been used for ground-penetrating radar (GPR) applications (Sato et al. 2003), through-wall detection system for life detection and monitoring (Elboushi et al. 2013a, b; Mochizuki et al. 2007; Yemelyanov et al. 2009) and early breast cancer detection (Zhou 2010; Mustafa et al. 2013).

Recently, many research efforts have been applied for developing modern microwave imaging systems to work in civil engineering applications (Kharkovsky and Zoughi 2007). Traditionally microwave imaging is performed by raster scanning a wideband antenna across a two-dimensional (2D) grid forming a synthetic aperture. The collected reflected data is then processed by a synthetic aperture-focusing algorithm to produce a 2D or 3D image of the sample under test. In this and many other applications small and directional UWB antennas are desired. However, a few UWB antennas have been reported. Most of the small UWB antennas are omnidirectional

M. Moosazadeh, *Antipodal Vivaldi Antennas for Microwave Imaging of Construction Materials and Structures*,
https://doi.org/10.1007/978-3-030-05566-0_1

in radiation characteristics. One of the widely used directional UWB antennas is a tapered slot antenna which can provide high directivity.

1.2 Research Objectives

The main aim of this research is to develop and apply novel UWB AVAs for high-resolution detection of defects and damages in composite construction materials and structures using their microwave and millimetre wave imaging. The challenge of applying UWB microwave technique is that it is dependent on the operating frequency utilised for specified material under test. Therefore, the objectives of this research are as follows:

1. To develop a small UWB antenna at a frequency range from 5 to 50 GHz for microwave and millimetre wave imaging of a wide range of low-loss construction materials
2. To design a small UWB antenna operating for microwave and millimetre wave imaging of low-loss and high-loss materials for the purpose of detection of surface damages of concrete under low-loss materials
3. To develop a UWB antenna at a frequency range from 2 to 27 GHz for microwave imaging of low-loss and high-loss materials such as concrete structures and layered structures for the purpose of detection of cavities inside concrete
4. To design a miniaturised UWB antenna at a frequency range from 1 to 30 GHz for microwave imaging of low-loss and high-loss materials such as concrete and layered structures for the purpose of detection of defects and structural features inside concrete
5. To develop a UWB antenna operating at lower frequency ~1 GHz and high gain at low frequencies for microwave imaging of low-loss and high-loss materials such as concrete and layered structures for the purpose of detection of void inside concrete structures.

1.3 Book Organisation

This book is organised into eight chapters as follows:

Chapter 1 includes a brief background, advantages and applications of the microwave imaging technology.

Chapter 2 provides literature review of the NDT&E methods and microwave and millimetre wave imaging techniques. A detailed discussion and challenges for antennas are presented.

Chapter 3 describes a design of a small and compact elliptically tapered AVA with wide operating bandwidth (5–50 GHz) for the application in UWB imaging system. Lower end of frequency band limitation for $S_{11} < -10$ dB has been extended by rectangular slits at sun-shaped configuration. The applied rectangular slits enhance antenna gain up to 5.7 dB at the lower frequencies. A half elliptical shaped dielectric lens as extension of antenna substrate is employed to enhance antenna gain at the higher frequencies, and as a result the antenna gain is 14 dB at 50 GHz. In addition, the proposed antenna demonstrates reduced E-plane tilt of beam, e.g. $-1°$ at 50 GHz, and lower side lobe level, cross-polarisation (< -22 dB), E-plane half power beam width (e.g. 12.2° at 50 GHz) and improved F-to-B ratio. Capability of the proposed AVA for a high-range resolution imaging of samples made of construction materials is demonstrated.

Chapter 4 presents a compact AVA with high gain at low frequencies and F-to-B ratio, lower operating frequency (3.4 GHz) and wide operating bandwidth from 3.4 to 40 GHz. It includes periodic slit-edge technique and trapezoid-shaped dielectric lens as an extension of substrate. Applicability of the proposed AVA for UWB imaging of construction materials and structures for NDT&E is demonstrated using three different samples including a concrete block with plasterboard sheets to highlight the high-range resolution of the proposed antenna. In the first sample, capability of the proposed antenna for the detection and evaluation of targets such as metal discs placed at different heights on foam substrate is demonstrated. In the second sample, a practical sample that is based on cubic concrete block was prepared and imaged. A groove was made on the top surface of the block and one steel rod was inserted in the groove. In the third sample, top of the cubic concrete block was covered by plasterboard sheets and an air gap was created by removing one sheet. Third sample had a cavity in a composite structure and imaging results are provided.

Chapter 5 presents a novel technique for design of AVA in order to lower the low end of frequency band to be applied for imaging of concrete-based specimens. To extend the low end of frequency band of the conventional antenna, inner edges of the top and bottom radiators of the antenna have been bent. To enhance gain at lower frequencies, regular triangular slit technique was applied to the antenna. Finally, applicability of the proposed antenna for UWB imaging of construction materials and structures for non-destructive testing and evaluation of concrete-based specimens was demonstrated to highlight the capability of the proposed antenna with the imaging system for the detection of flaws such as void inside construction materials with high-range resolution (e.g. a few millimetres in concrete) and deep penetration depth and propagation path (e.g. more than 200 mm in two ways in concrete).

Chapter 6 presents a miniaturised AVA with a minimum operating frequency of 1 GHz and wide operating frequency band from 1 to 30 GHz. It includes bending technique on inner edges of radiators to lower operating frequency band to 1 GHz. Regular slits have been employed in edges of radiators to enhance gain at low frequencies. The half-elliptical shaped dielectric lens as an extension of substrate has been used to further enhance antenna gain. The proposed antenna has been applied for microwave imaging of high-loss materials such as concrete. The capability of the proposed antenna for UWB imaging has been demonstrated for the detection of voids inside concrete.

Chapter 7 presents a design of AVA with high gain at low frequencies. It consists of comb-shaped slits in edges of the top and bottom radiators to lower cut-off frequency of conventional antipodal Vivaldi antenna and to enhance antenna gain at low frequencies. The proposed antenna has impedance bandwidth of 1.65–18 GHz, high gain at low frequencies (6.7 dB at 1.65 GHz and 9.1 dB at 2 GHz) and F-to-B ratio of 42 dB at 13.5 GHz. The capability of the proposed AVA for UWB imaging for the purpose of the detection of voids inside a concrete beam is demonstrated. Extended results of microwave imaging of voids inside concrete at different standoff distance between the proposed antenna and the surface of concrete are also provided.

Chapter 8 provides concluding remarks of the book and some possible suggestions for future works.

References

Elboushi, A., Joanes, D., Derbas, M., Khaled, S., Zafar, A., Attabibi, S., & Sebak, A. (2013a). *Design of UWB antenna array for through-wall detection system.* In 2013 IEEE Symposium on Wireless Technology and Applications (ISWTA). IEEE, pp. 349–354.

Elboushi, A., Sebak, A., & Denidni, T. (2013b). Through wall gap detection using monostatic radar. *Applied Computational Electromagnetics Society Journal, 28*, 411–418.

Kharkovsky, S., & Zoughi, R. (2007). Microwave and millimeter wave nondestructive testing and evaluation-overview and recent advances. *IEEE Instrumentation and Measurement Magazine, 10*, 26–38.

Kharkovsky, S., Case, J. T., Ghasr, M. T., Zoughi, R., Bae, S. W., & Belarbi, A. (2012). Application of microwave 3D SAR imaging technique for evaluation of corrosion in steel rebars embedded in cement-based structures. In *Review of progress in quantitative nondestructive evaluation* (Vol. 31, pp. 1516–1523). Melville: AIP.

Mochizuki, S., Togo, H., Kukutsu, N., & Nagatsuma, T. (2007). *High-resolution SAR imaging system using tapered slot antenna with orthogonally anisotropic directivity.* In 2007 Asia-Pacific Conference on Applied Electromagnetics. APACE 2007. IEEE, pp. 1–4.

Mustafa, S., Mohammed, B., & Abbosh, A. (2013). Novel pre-processing techniques for accurate microwave imaging of human brain. *IEEE Antennas and Wireless Propagation Letters, 12*, 460–463.

Roqueta, G., Jofre, L., & Feng, M. Q. (2012). Analysis of the electromagnetic signature of reinforced concrete structures for nondestructive evaluation of corrosion damage. *IEEE Transactions on Instrumentation and Measurement, 61*, 1090–1098.

Sato, M., Fang, G., & Zeng, Z. (2003). *Landmine detection by a broadband GPR system.* In 2003 IEEE International Geoscience and Remote Sensing Symposium, 2003. IGARSS'03. Proceedings. IEEE, pp. 758–760.

Yemelyanov, K. M., Engheta, N., Hoorfar, A., & Mcvay, J. A. (2009). Adaptive polarization contrast techniques for through-wall microwave imaging applications. *IEEE Transactions on Geoscience and Remote Sensing, 47*, 1362–1374.

Zhou, Y. (2010). Microwave imaging based on wideband range profiles. *Progress in Electromagnetics Research Letters, 19*, 57–65.

Chapter 2
Literature Review

2.1 Introduction

NDT&E is the science of evaluating different properties of a material without compromising its utility (Shull 2002). Usually, NDT&E is carried out with an interrogating electric, magnetic or acoustic signal. The signals interact with the materials, and the responses give properties of the material under test. The majority of well-established NDT&E methods for a variety of applications are eddy current, visual testing and ultrasound methods. However, each of the aforementioned techniques has its own merit, drawbacks and realm of applicability.

Ultrasonic techniques are employed for data acquisition and they incorporate ultrasonic instrumentation and computer controls; however, ultrasonic techniques have drawback for coupling medium requirement (Shull 2002). On the other hand, ultrasonic signals may not penetrate into highly porous materials and as a result they cannot interact with their inner structure. Eddy current methods do not work well when inspecting low-loss dielectric materials in which eddy currents cannot be induced (Shull 2002).

Materials technology has produced stiffer, lighter and more durable electrically insulating composites which are replacing metals in certain applications. These composites require alternative inspection techniques because the standard NDT&E methods may not be capable of inspecting them.

In some cases, microwave and millimetre wave NDT&E techniques may be the only viable solution and it presents comprehensive picture of material under test (Kharkovsky and Zoughi 2007).

The well-established microwave frequency spectrum spans from about 300 MHz to 30 GHz and the corresponding wavelength range is 1000–10 mm, while the frequency spectrum of 30–300 GHz is related to the millimetre wave spectrum and the corresponding wavelength range is 10–1 mm (Pozar 2012).

M. Moosazadeh, *Antipodal Vivaldi Antennas for Microwave Imaging of Construction Materials and Structures*,
https://doi.org/10.1007/978-3-030-05566-0_2

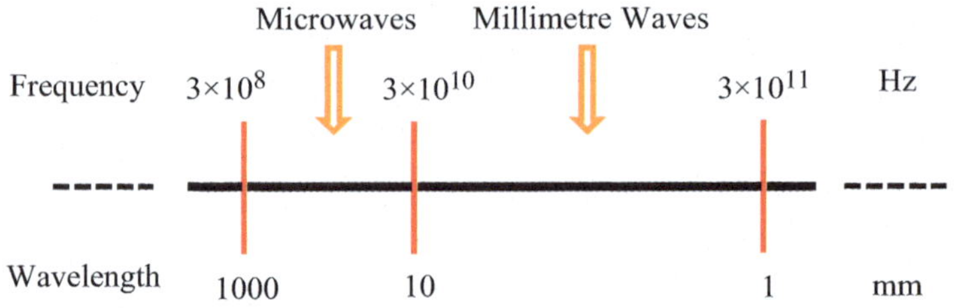

Fig. 2.1 Microwave and millimetre wave frequency ranges and corresponded wavelengths

Figure 2.1 displays the frequency range and wavelengths pertaining to the microwave and millimetre waves. Interrogating signals at microwave and millimetre wave frequencies can readily penetrate into dielectric materials and are able to interact with their interior structure.

The utility of microwave and millimetre wave NDT&E has been applied to the following diverse applications:

- Dielectric material characterisation using a diverse array of techniques and for a wide range of applications including dielectric property characterisation, dielectric mixture constituent determination, porosity evaluation in polymers, moisture measurements, cure monitoring of resin binders, rubber products and cement-based materials (Bois et al. 1999; Kharkovsky et al. 2002; Hughes and Zoughi 2005)
- Fatigue crack detection and sizing in metal surfaces, including those under paint and thin dielectric coatings using several unique approaches and detection of other surface anomalies in metals (Yeh and Zoughi 1994)
- Detection and evaluation of corrosion under paint and composite laminates, including detection of corrosion precursor pitting (Ghasr et al. 2005, 2006)
- Near-field, focused and synthetic aperture radar microwave and millimetre wave imaging for flaw detection and evaluation in various composite structures (Shrestha et al. 2005; Kharkovsky et al. 2006)

The dielectric properties of a material under test can be described by its effective complex dielectric permittivity and loss tangent (Pozar 2012) as

$$\epsilon = \left(\epsilon' - j\epsilon''_{\text{eff}}\right), \tag{2.1}$$

$$\tan\delta = \frac{\epsilon''_{\text{eff}}}{\epsilon'} = \frac{\epsilon'' + \sigma/\omega}{\epsilon'}, \tag{2.2}$$

where $\epsilon' = \epsilon_o\epsilon_r$ is the real part of the effective complex permittivity of the material, ϵ_0 is the permittivity of free space, ϵ_r is the relative dielectric constant, $\epsilon''_{\text{eff}} = \epsilon'' + \sigma/\omega$ is the imaginary part of the effective complex permittivity of the material, $\epsilon'' = \epsilon_0\epsilon''_r$ is the effective loss factor of the material, ϵ''_r is the relative loss factor which accounts for loss in the material in the form of heat due to the vibrating dipole moments, σ is the conductivity of the material (S/m) and ω is the angular frequency (radian).

2.2 Spot Measurement Techniques

A comprehensive study of the electromagnetic reflections that take place inside concrete structures with corroded rebars has been performed and detailed in Roqueta et al. (2012). A network analyser is utilised to generate excitation signal and for the signal acquisition at the frequency domain. For this purpose, UWB planar elliptical shaped monopole antennas were used. Polarisation of these antenna should be parallel to the steel rebar in order to maximise the sensitivity to steel rebar corrosion. The antennas are located at few centimetres from the concrete specimen surface. Three measurement setups were applied: (a) through-transmission, (b) monostatic-reflection and (c) bistatic set-ups to determine the best option for corrosion characterisation.

2.3 Microwave and Millimetre Wave Imaging Techniques

One of the applications of microwave imaging is the non-destructive detection and evaluation of defects, damages and structural features inside construction materials including plastics, wood, rubber, plasterboard and concrete that have different electrical properties (i.e. complex dielectric permittivity and conductivity) (Sagnard and Zein 2005; Muqaibel and Safaai-Jazi 2003). It means, for example, that they can be low loss (e.g. plastics) and high loss (e.g. wet concrete), i.e. they have different reflection properties, transmission properties and depth of penetration.

To perform a three-dimensional (3-D) imaging of a material under test, a two-dimensional (2-D) aperture has to be sampled with a broadband measurement signal at each selected transmit-receive combination. The spatial extension of the aperture determines the lateral resolution $\delta_{x,\,y}$, in material under investigation, given by (Ahmed et al. 2012)

$$\delta_{x,y} = \frac{\lambda}{D_{x,y}} L, \tag{2.3}$$

where $D_{x,\,y}$ denotes the length of the aperture in the corresponding direction, λ the wavelength and L the distance between object and aperture. The resolution δ_z in range direction in material under test is determined approximately by the signal bandwidth B, and thus given by

$$\delta_z \approx \frac{c}{2B\sqrt{\varepsilon_r}}, \tag{2.4}$$

where c is the speed of light in free space.

Accordingly, a large signal bandwidth B results in an equivalent short pulse duration and hence in a high-range resolution; for instance the value of range resolution for dielectric constant of 4 and bandwidth of 8 GHz, 16 GHz and 20 GHz will be 9.37 mm, 4.68 mm and 3.75 mm, respectively. In practice, the bandwidth is often limited by the employed antennas and semiconductor components, e.g. oscillators, mixers and amplifiers (Ahmed et al. 2012).

The penetration depth in material under investigation can be evaluated using the following equations (Shaari et al. 2004):

$$\delta_s = \frac{1}{\alpha} \tag{2.5}$$

$$\alpha = \left(\frac{\omega}{c}\right)\left[\left(\frac{\varepsilon_r'}{2}\right)\left(\sqrt{1+\left(\frac{\varepsilon_r''}{\varepsilon_r'}\right)^2}-1\right)\right]^{1/2} \tag{2.6}$$

where α is the attenuation coefficient, ω is $2\pi f$ and f is the frequency.

Microwave signals are capable to penetrate inside materials such as composites, concrete, foam and wood and interact with interior structures. It is important to note that signals at microwave and millimetre wave frequencies are nonionising (Kharkovsky and Zoughi 2007). In microwave and millimetre wave imaging, a microwave signal is transmitted to the targeted object and backscattered signals from the object under test are processed to reconstruct images of the targeted objects (Pastorino 2010; Akinci et al. 2015).

Over the past few decades, significant progress has been achieved in UWB imaging systems and image algorithm development, which increasingly promotes this technique for applications such as medical applications (Klemm et al. 2010), security scanning (Sheen et al. 2001; Zhuge and Yarovoy 2011; Ahmed et al. 2012), through-wall imaging (Yang and Fathy 2009; Lubecke et al. 2007; Ralston et al. 2010) and NDT&E (Kharkovsky and Zoughi 2007; Ghasr et al. 2013).

2.4 Antennas for Microwave Imaging Applications

Antennas play a crucial role in obtaining desired performance of the microwave and millimetre wave imaging techniques. Microwave imaging is being accomplished by scanning an antenna or antenna array over an object (Hatfield et al. 2013; Yarovoy et al. 2007; Zhuge and Yarovoy 2010; Kharkovsky et al. 2013). The success of UWB microwave imaging in these applications stems from the ability of microwave signals to penetrate non-metallic structures and safely provide images of their interior. Furthermore, utilising imaging algorithms such as those derived from synthetic aperture focusing (SAF) results in high-resolution 2-D and 3-D images (Case et al. 2011; Ghasr et al. 2012).

In general, four principal requirements for the microwave techniques should be employed for imaging applications, namely they should provide (1) directional radiation, (2) high spatial and range resolution, (3) desired penetration depth of electromagnetic waves inside materials and structures under test and (4) portability, i.e. small size of the microwave devices is essential (Pastorino 2010; Ahmed et al. 2012; Sagnard and Zein 2005). Therefore, to satisfy the four principal requirements for non-destructive testing and evaluation of a variety of construction materials and composites, advanced microwave imaging methods and techniques are required. For this purpose, UWB techniques (Zwick et al. 2013) may be one of the options. Although the achievements are impressive, challenges of UWB imaging still exist that must be solved before its wide adoption of use in practice. For instance, the development and modification and/or improvement of UWB antennas are needed for the NDT&E of construction composite materials and structures. In the next section, omnidirectional UWB antennas and directional UWB antennas such as open-ended waveguide, horn, spiral and Vivaldi antennas are considered.

2.4.1 *Omnidirectional UWB Antennas*

As the demand of compact, smart and multifunctional antennas for modern applications including communication, civil engineering and biomedical applications increases, wideband antennas attract more and more interests in the academic field (Porcino and Hirt 2003; Leeper 2003; Rodenbeck et al. 2005). UWB antennas manufactured using printed circuit board (PCB) technologies are relatively small, light and inexpensive. They may provide an operating frequency band that covers a few waveguide frequency bands. PCB-based UWB antennas are used in UWB communication systems operating at the frequency range from 3.1 to 10.6 GHz according to Federal Communication Commission (FCC). They have wide frequency bandwidth and symmetrical beamwidths. Many methods such as the truncated slot/notch on the structure antenna have been reported for increasing the frequency range but each of them has its own merits and drawbacks. Figure 2.2 shows a configuration of the reported UWB antenna with the checkered-shaped patch.

The majority of the UWB antennas exhibit omnidirectional radiation patterns, low gain (i.e. typically 6 dBi) and an impulse response with significant distortion. Thus, these kinds of UWB compact antennas are suitable for the short-range indoor and outdoor communication.

2.4.2 *Directional UWB Antennas*

- **Open-ended waveguide antennas**

Open-ended waveguide antennas/probes have been widely used for microwave and millimetre wave NDT&E near-field techniques (Kharkovsky and Zoughi 2007).

Advantages of these techniques include the fact that they are noncontact, one sided, compact, real time, robust, rugged and relatively inexpensive. Open-ended rectangular waveguide antennas and adaptors at R- (1.7–2.6 GHz), S- (2.6–3.95 GHz), G- (3.95–5.85 GHz) and X-band (8.2–12.4 GHz) are shown in Fig. 2.3. Disadvantages of these antennas include their limited frequency bandwidth and a relatively low realised gain due to a high reflection from the aperture as a result of impedance mismatching between a waveguide and free space.

- **Vivaldi and antipodal Vivaldi antennas**

The prominent option for microwave and millimetre wave imaging applications is the use of Vivaldi antenna which was invented by Gibson in 1979 (Gibson 1979) since it has wide bandwidth, directional radiation pattern, high gain, lightweight, planar structure, low profile and easy integration with devices. However, there are some challenges concerning Vivaldi antennas including complexity of feeding struc-

Fig. 2.2 Photograph of the UWB antenna with the checkered-shaped patch (Moosazadeh et al. 2010)

Fig. 2.3 Microwave open-ended waveguides and adapters at R- (1.7–2.6 GHz), S- (2.6–3.95 GHz), G- (3.95–5.85 GHz) and X-band (8.2–12.4 GHz)

ture, large size, limitation on low end of frequency band, low gain at low and/or high frequencies, and instability of directivity in radiation patterns, i.e. E-plane tilt of beam especially at high frequencies. Feeding element of the Vivaldi antennas is a microstrip to slot-line transition; however, the impedance bandwidth of the Vivaldi antenna is limited because of complex feed design (Ebnabbasi et al. 2013; Shuppert 1988; Greenberg et al. 2003). Figure 2.4 shows the geometry of the Vivaldi antenna.

In 1988, Gazit (Gazit 1988) introduced antipodal Vivaldi antenna (AVA) with two radiators on opposite sides of substrate, namely top and bottom layers. Its feed transition element is a microstrip to symmetric double-side slot line, resulting in theoretically infinite bandwidth. Figure 2.5 shows the sketch of the conventional AVA (CAVA).

Various designs of the CAVAs have been presented in the literature. In Qing et al. (2008) parametric study of the conventional AVA (CAVA) has been documented. However, gain of the CAVAs is low and impedance bandwidth of the CAVAs is limited at the low end of frequency band.

To improve radiation characteristics and extend low end of frequency band of the CAVAs, various kinds of the slits/slots have been used and reported in the literature. Circular-shape-load and slot-load AVA is presented in Bai et al. (2011) to extend further the low end of frequency band. An exponential shaped slot on edges of AVA is presented in De Oliveira et al. (2015) to improve radiation characteristics and extend limitation on the low end of frequency band. In Natarajan et al. (2015), structural modifications along with L-shaped slot in top and bottom radiators have been used in order to reduce lower operating frequency. In Kanjaa et al. (2015), exponential slots in top and bottom radiators of the CAVA have been used to improve impedance matching at low frequencies. In Natarajan et al. (2016), impedance bandwidth is extended by combination of two AVAs (AVA with upper petal and AVA with lower petal). In Ying and Dou (2013), U-shaped tapered slot edge is employed to lower low end of frequency band and improve radiation characteristics at low frequencies. However, these antennas have low gain in the entire band or, in some cases, at high frequencies.

To increase antenna gain at high frequencies, dielectric directors and lenses have been employed to AVAs as extension of antenna's substrate to direct the electromagnetic waves toward the endfire direction. For instance, the use of loaded dielectric

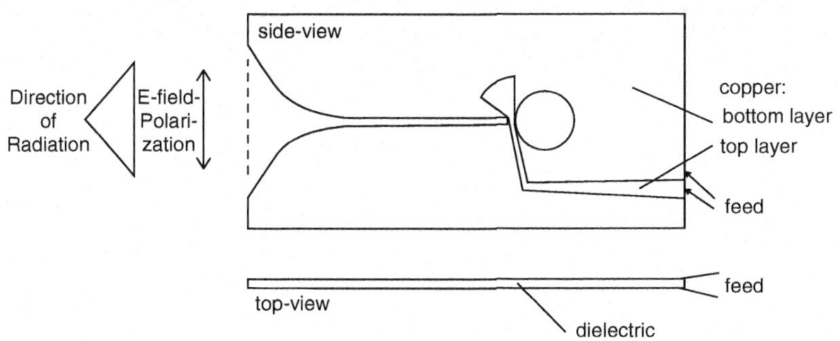

Fig. 2.4 Sketch of a Vivaldi antenna

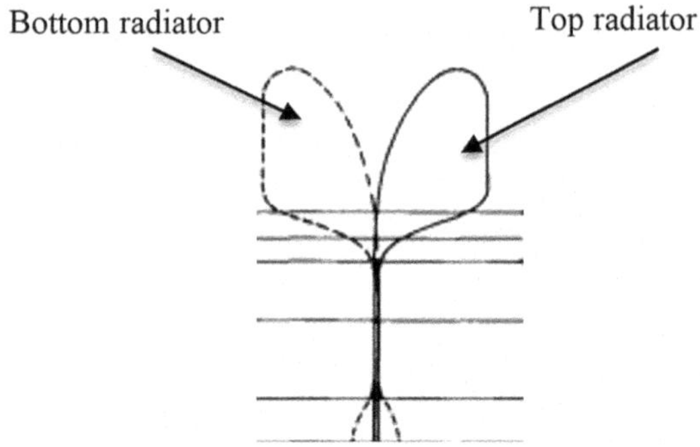

Fig. 2.5 Sketch of the conventional antipodal Vivaldi antenna

lens (Li et al. 2016), half elliptical shaped dielectric lens (Molaei et al. 2014; Ashraf et al. 2015), dielectric directors (Bourqui et al. 2010; Juan et al. 2013) and parasitic elliptical patch (Nassar and Weller 2015) has been reported in the literature.

However, reported AVAs with dielectric directors and lenses have its own drawbacks. For instance, antenna gain is low at their highest frequency of bandwidth and E-plane radiation patterns are tilted at high frequencies. In some cases, the dielectric directors have been used with higher dielectric constant than that of the antenna substrate to direct most of the energy toward the aperture centre (Juan et al. 2013; Bourqui et al. 2010).

2.5 Summary

In this chapter, literature review of different types of antennas has been presented. Of particular interest for civil engineering applications is the use of UWB measurement techniques that can provide high-range resolutions for the detection of defects and damages in construction materials and structures. The most challenging part of these UWB techniques is the design of UWB antennas with high gain at low and/or high frequencies, high F-to-B ratio, small size and low profile. It is required that the UWB antennas have lower operating bandwidth and wide frequency band to provide deep penetration depth and high-range resolution, respectively. In the following chapters, five UWB AVAs with different characteristics will be developed and presented. The performance of designed UWB AVAs in free space will be investigated and then they will be employed for microwave imaging of construction materials and structures.

References

Ahmed, S. S., Schiessl, A., Gumbmann, F., Tiebout, M., Methfessel, S., & Schmidt, L. (2012). Advanced microwave imaging. *IEEE Microwave Magazine, 13*, 26–43.

Akinci, M. N., Caglayan, T., Ozgur, S., Alkasi, U., Ahmadzay, H., Abbak, M., Cayoren, M., & Akduman, I. (2015). Qualitative microwave imaging with scattering parameters measurements. *IEEE Transactions on Microwave Theory and Techniques, 63*, 2730–2740.

Ashraf, M., Haraz, O., Sebak, A., & Alshebeili, S. (2015). *Wideband compact Vivaldi antenna loaded with dielectric lens for millimeter-wave applications*. In 2015 IEEE International Symposium on Antennas and Propagation & USNC/URSI National Radio Science Meeting. IEEE, pp. 2061–2062.

Bai, J., Shi, S., & Prather, D. W. (2011). Modified compact antipodal Vivaldi antenna for 4–50-GHz UWB application. *IEEE Transactions on Microwave Theory and Techniques, 59*, 1051–1057.

Bois, K. J., Handjojo, L. F., Benally, A. D., Mubarak, K., & Zoughi, R. (1999). Dielectric plug-loaded two-port transmission line measurement technique for dielectric property characterization of granular and liquid materials. *IEEE Transactions on Instrumentation and Measurement, 48*, 1141–1148.

Bourqui, J., Okoniewski, M., & Fear, E. C. (2010). Balanced antipodal Vivaldi antenna with dielectric director for near-field microwave imaging. *IEEE Transactions on Antennas and Propagation, 58*, 2318–2326.

Case, J. T., Ghasr, M. T., & Zoughi, R. (2011). Optimum two-dimensional uniform spatial sampling for microwave SAR-based NDE imaging systems. *IEEE Transactions on Instrumentation and Measurement, 60*, 3806–3815.

De Oliveira, A. M., Perotoni, M. B., Kofuji, S. T., & Justo, J. F. (2015). A palm tree antipodal Vivaldi antenna with exponential slot edge for improved radiation pattern. *IEEE Antennas and Wireless Propagation Letters, 14*, 1334–1337.

Ebnabbasi, K., Sczyslo, S., & Mohebbi, M. (2013). UWB performance of coplanar tapered slot antennas. *IEEE Antennas and Wireless Propagation Letters, 12*, 749–752.

Gazit, E. (1988). *Improved design of the Vivaldi antenna*. In IEE Proceedings H (Microwaves, Antennas and Propagation). IET, pp. 89–92.

Ghasr, M. T., Kharkovsky, S., Zoughi, R., & Austin, R. (2005). Comparison of near-field millimeter-wave probes for detecting corrosion precursor pitting under paint. *IEEE Transactions on Instrumentation and Measurement, 54*, 1497–1504.

Ghasr, M. T., Carroll, B., Kharkovsky, S., Austin, R., & Zoughi, R. (2006). Millimeter wave differential probe for nondestructive detection of corrosion precursor pitting. *IEEE Transactions on Instrumentation and Measurement, 55*, 1620–1627.

Ghasr, M. T., Abou-Khousa, M. A., Kharkovsky, S., Zoughi, R., & Pommerenke, D. (2012). Portable real-time microwave camera at 24 GHz. *IEEE Transactions on Antennas and Propagation, 60*, 1114–1125.

Ghasr, M. T., Kharkovsky, S., Bohnert, R., Hirst, B., & Zoughi, R. (2013). 30 GHz linear high-resolution and rapid millimeter wave imaging system for NDE. *IEEE Transactions on Antennas and Propagation, 61*, 4733–4740.

Gibson, P. (1979). *The Vivaldi aerial*. In 9th European Microwave Conference, 1979. IEEE, pp. 101–105.

Greenberg, M. C., Virga, K. L., & Hammond, C. L. (2003). Performance characteristics of the dual exponentially tapered slot antenna (DETSA) for wireless communications applications. *IEEE Transactions on Vehicular Technology, 52*, 305–312.

Hatfield, S., Hillstrom, M., Schultz, D., Werckmann, T., Ghasr, M., & Donnell, K. (2013). *UWB microwave imaging array for nondestructive testing applications*. In 2013 IEEE International Instrumentation and Measurement Technology Conference (I2MTC). IEEE, pp. 1502–1506.

Hughes, D., & Zoughi, R. (2005). A novel method for determination of dielectric properties of materials using a combined embedded modulated scattering and near-field microwave techniques-part I: Forward model. *IEEE Transactions on Instrumentation and Measurement, 54*, 2389–2397.

Juan, L., Guang, F., Lin, Y., & Demin, F. (2013). A modified balanced antipodal Vivaldi antenna with improved radiation characteristics. *Microwave and Optical Technology Letters, 55*, 1321–1325.

Kanjaa, M., Mrabet, O. E., Khalladi, M., & Essaaidi, M. (2015). *Exponentially tapered antipodal Vivaldi antenna for breast cancer detection*. In 2015 IEEE 15th Mediterranean Microwave Symposium (MMS). IEEE, pp. 1–3.

Kharkovsky, S., & Zoughi, R. (2007). Microwave and millimeter wave nondestructive testing and evaluation-overview and recent advances. *IEEE Instrumentation and Measurement Magazine, 10*, 26–38.

Kharkovsky, S. N., Akay, M. F., Hasar, U. C., & ATIS, C. D. (2002). Measurement and monitoring of microwave reflection and transmission properties of cement-based specimens. *IEEE Transactions on Instrumentation and Measurement, 51*, 1210–1218.

Kharkovsky, S., Case, J. T., Abou-Khousa, M. A., Zoughi, R., & Hepburn, F. L. (2006). Millimeter-wave detection of localized anomalies in the space shuttle external fuel tank insulating foam. *IEEE Transactions on Instrumentation and Measurement, 55*, 1250–1257.

Kharkovsky, S., Ghasr, M. T., Kam, K., Abou-Khousa, M. A., & Zoughi, R. (2013). Out-of plane fed elliptical slot array for microwave imaging. *IEEE Transactions on Antennas and Propagation, 61*, 5311–5314.

Klemm, M., Leendertz, J., Gibbins, D., Craddock, I. J., Preece, A., & Benjamin, R. (2010). Microwave radar-based differential breast cancer imaging: Imaging in homogeneous breast phantoms and low contrast scenarios. *IEEE Transactions on Antennas and Propagation, 58*, 2337–2344.

Leeper, D. G. (2003). *Ultrawideband-the next step in short-range wireless*. In IEEE Radio Frequency Integrated Circuits (RFIC) Symposium, 2003. IEEE, pp. 493–496.

Li, L., Xia, X., Liu, Y., & Yang, T. (2016). Wideband balanced antipodal Vivaldi antenna with enhanced radiation parameters. *Progress in Electromagnetics Research C, 66*, 163–171.

Lubecke, V. M., Boric-Lubecke, O., Host-Madsen, A., & Fathy, A. E. (2007). *Through-the-wall radar life detection and monitoring*. In 2007 IEEE/MTT-S International Microwave Symposium. IEEE, pp. 769–772.

Molaei, A., Kaboli, M., Mirtaheri, S. A., & Abrishamian, M. S. (2014). Dielectric lens balanced antipodal Vivaldi antenna with low cross-polarisation for ultra-wideband applications. *IET Microwaves, Antennas and Propagation, 8*, 1137–1142.

Moosazadeh, M., Ghobadi, C., & Dousti, M. (2010). Small monopole antenna with checkered-shaped patch for UWB application. *IEEE Antennas and Wireless Propagation Letters, 9*, 1014–1017.

Muqaibel, A. H., & Safaai-Jazi, A. (2003). A new formulation for characterization of materials based on measured insertion transfer function. *IEEE Transactions on Microwave Theory and Techniques, 51*, 1946–1951.

Nassar, I. T., & Weller, T. M. (2015). A novel method for improving antipodal Vivaldi antenna performance. *IEEE Transactions on Antennas and Propagation, 63*, 3321–3324.

Natarajan, R., George, J. V., Kanagasabai, M., & Shrivastav, A. K. (2015). A compact antipodal Vivaldi antenna for UWB applications. *IEEE Antennas and Wireless Propagation Letters, 14*, 1557–1560.

Natarajan, R., George, J. V., Kanagasabai, M., Lawrance, L., Moorthy, B., Rajendran, D. B., & Alsath, M. G. N. (2016). Modified antipodal Vivaldi antenna for ultra-wideband communications. *IET Microwaves, Antennas and Propagdtion, 10*, 401–405.

Pastorino, M. (2010). *Microwave imaging*. Hoboken, NJ: Wiley.

Porcino, D., & Hirt, W. (2003). Ultra-wideband radio technology: Potential and challenges ahead. *IEEE Communications Magazine, 41*, 66–74.

Pozar, D. M. (2012). *Microwave engineering*. New York: Addison-Wesley.

Qing, X., Chen, Z. N., & Chia, M. Y. W. (2008). Parametric study of ultra-wideband dual elliptically tapered antipodal slot antenna. *International Journal of Antennas and Propagation, 2008*, 267197.

Ralston, T. S., Charvat, G. L., & Peabody, J. E. (2010). *Real-time through-wall imaging using an ultrawideband multiple-input multiple-output (MIMO) phased array radar system*. In 2010 IEEE International Symposium on Phased Array Systems and Technology (ARRAY). IEEE, pp. 551–558.

Rodenbeck, C. T., Kim, S.-G., Tu, W.-H., Coutant, M. R., Hong, S., Li, M., & Chang, K. (2005). Ultra-wideband low-cost phased-array radars. *IEEE Transactions on Microwave Theory and Techniques, 53*, 3697–3703.

Roqueta, G., Jofre, L., & Feng, M. Q. (2012). Analysis of the electromagnetic signature of reinforced concrete structures for nondestructive evaluation of corrosion damage. *IEEE Transactions on Instrumentation and Measurement, 61*, 1090–1098.

Sagnard, F., & Zein, G. E. (2005). In situ characterization of building materials for propagation modeling: Frequency and time responses. *IEEE Transactions on Antennas and Propagation, 53*, 3166–3173.

Shaari, A., Millard, S., & Bungey, J. (2004). Modelling the propagation of a radar signal through concrete as a low-pass filter. *NDT and E International, 37*, 237–242.

Sheen, D. M., McMakin, D. L., & Hall, T. E. (2001). Three-dimensional millimeter-wave imaging for concealed weapon detection. *IEEE Transactions on Microwave Theory and Techniques, 49*, 1581–1592.

Shrestha, S., Kharkovsky, S., Zoughi, R., & Hepburn, F. (2005). Microwave and millimeter wave nondestructive testing of the space shuttle external tank insulating foam. *Materials Evaluation, 63*, 339–344.

Shull, P. J. (2002). *Nondestructive evaluation: Theory, techniques, and applications*. New York: CRC press.

Shuppert, B. (1988). Microstrip/slot-line transitions: Modeling and experimental investigation. *IEEE Transactions on Microwave Theory and Techniques, 36*, 1272–1282.

Yang, Y., & Fathy, A. E. (2009). Development and implementation of a real-time see-through-wall radar system based on FPGA. *IEEE Transactions on Geoscience and Remote Sensing, 47*, 1270–1280.

Yarovoy, A. G., Savelyev, T. G., Aubry, P. J., Lys, P. E., & Ligthart, L. P. (2007). UWB array-based sensor for near-field imaging. *IEEE Transactions on Microwave Theory and Techniques, 55*, 1288–1295.

Yeh, C.-Y., & Zoughi, R. (1994). A novel microwave method for detection of long surface cracks in metals. *IEEE Transactions on Instrumentation and Measurement, 43*, 719–725.

Ying, Q., & Dou, W. (2013). *Simulation of two compact antipodal Vivaldi antennas with Radiation Characteristics enhancement*. In 2013 Proceedings of the International Symposium on Antennas & Propagation (ISAP). IEEE, pp. 523–526.

Zhuge, X., & Yarovoy, A. (2010). *Design of low profile antipodal Vivaldi antenna for ultrawide-band near-field imaging*. In Proceedings of the Fourth European Conference on Antennas and Propagation. IEEE, pp. 1–5.

Zhuge, X., & Yarovoy, A. G. (2011). A sparse aperture MIMO-SAR-based UWB imaging system for concealed weapon detection. *IEEE Transactions on Geoscience and Remote Sensing, 49*, 509–518.

Zwick, T., Wiesbeck, W., Timmermann, J., & Adamiuk, G. (2013). *Ultra-wideband RF system engineering*. Cambridge: Cambridge University Press.

Chapter 3
Small UWB Antipodal Vivaldi Antenna with Improved Radiation Characteristics for Microwave and Millimetre Wave Imaging Applications

3.1 Introduction

Microwave and millimetre wave imaging of materials and structures is a very useful and effective technique with advantages including its nonionisation nature and relatively low cost and low complexity (Pastorino 2010; Ahmed et al. 2012; Zoughi 2000). For example, relatively simple single-frequency or narrow-frequency band microwave or millimetre wave imaging systems have been used for NDT&E of various metal, dielectric and composite materials (Kharkovsky and Zoughi 2007; Ghasr et al. 2012; Kharkovsky et al. 2012). It was shown that operating frequencies or frequency bands should be optimised for each type of materials under test. For instance, the inspection of spray-on-foam insulation required millimeter wave frequencies up to 150 GHz while NDT&E of the state of rebars in mortar was performed at microwave frequencies of X-band (8.2–12.4 GHz). It should be noted that there is a variety of construction materials with different dielectric properties such as glass, polymers, plasterboards and concrete. To provide NDT&E of these materials UWB techniques are desired. In addition, in many cases, small and directive UWB antennas are required to provide access to regions of interest of the structure under investigation by a single antenna or array of antennas.

In this chapter the development and application of a relatively small AVA for NDT&E of relatively low-loss construction materials such as construction foam and plasterboards with high-range resolution are presented. It is based on the design of a conventional AVA which size is 30 × 40 mm, and a desired frequency band from 5 to 50 GHz (i.e. its bandwidth is 45 GHz). To address the requirements of microwave and millimeter imaging of the construction materials, the proposed antenna should have the improved radiation characteristics. The improvement of characteristics includes higher gain at higher frequencies and front-to-back (F-to-B) ratio, narrower half-power beamwidth (HPBW), lower cross-polarisation and side lobe

M. Moosazadeh, *Antipodal Vivaldi Antennas for Microwave Imaging of Construction Materials and Structures*,
https://doi.org/10.1007/978-3-030-05566-0_3

levels and *E*-plane tilt of beam. The achievement of these characteristics in the UWB antennas with relatively small dimensions is the major challenge.

The chapter includes four sections. In the second section, the design of proposed antenna with rectangular slits at a sun-shaped configuration and a dielectric lens is proposed. Antenna performance in free space such as radiation characteristics and electric field distribution is provided in the third section. In the fourth section, simulation and measurement results for the proposed antenna irradiating samples made of construction materials are presented.

3.2 Proposed Antenna Configuration

A step-by-step modification procedure is employed to design the proposed antenna. First, conventional AVA (CAVA) is designed as a reference. Second, the modified AVA (MAVA) with rectangular slits at a sun-shaped configuration is introduced to extend lower end of frequency band limitation and increase antenna gain at lower frequencies. Third, the MAVA with a half elliptical shaped dielectric lens as an extension of the antenna substrate is designed (referred to as MAVA-HEDL) to feature high gain at higher frequencies, high F-to-B ratio, small side lobe level, narrow *E*-plane HPBW, lower cross-polarisation level and modification on *E*-plane tilt of beam.

3.2.1 Rectangular Slits at Sun-Shaped Configuration

The step-by-step modification procedure is demonstrated as shown in Fig. 3.1a–c, namely CAVA (Ant. 0), a rectangular slit at sun-shaped configuration AVA (SSCAVA) (Ant. 1) and MAVA (Ant. 2). The total size ($L_{SUB} \times W_{SUB}$) of the designed antennas is 30 × 40 mm. The antennas have been designed on Rogers RO4003C with relative permittivity of 3.38, loss tangent of 0.0027, thickness of 0.508 mm and copper thickness of 35 μm. The designed antennas are fed through microstrip line. The feed width is fixed on 1.19 mm to match with 50-Ω coaxial line. The top and bottom radiators are formed by the intersection of two ellipses for only one quadrant with major radii (W_{p1} and W_{p2}) and secondary radii (L_{p1} and L_{p2}). The ground plane is elliptically tapered from the intersection of a rectangle conductor with dimension of $W_g \times L_f$ and two anti-faced quarter ellipses with major radii of a_g and secondary radii of b_g as shown in Fig. 3.1d.

The simulated magnitude of reflection coefficient, S_{11} in dB, of the designed antennas at the whole frequency band is plotted in Fig. 3.2a. As shown in Fig. 3.2a, the behaviour of S_{11} for the all three antennas is similar with some differences at lower frequencies. In order to further emphasise on the magnitude of reflection coefficient especially at lower frequencies, frequency range from 9 to 11 GHz is selected and depicted in Fig. 3.2b. As shown in Fig. 3.2b, the cut-off frequency for the CAVA (Ant. 0) is 9.9 GHz. To further widen bandwidth, low end of frequency band limitation has been extended by rectangular slits at sun-shaped configuration designed at the edges of radiators by rectangular slits with length of L_{s1} and width of the W_{s1}, creating SSCAVA (i.e. Ant. 1). The rotation angle, α, of the slits is optimised at 20°. The L_{s1} is equal to 5.8 mm which is approximately $0.25\lambda / \sqrt{\varepsilon_r}$, where λ is the wavelength corresponding to 9 GHz. It can be seen from Fig. 3.2b that the low end of frequency band of the Ant. 1 is 5 GHz. However, it is observed that a poor impedance matching is occurred around 9–9.5 GHz. To improve impedance matching, a quarter of a sun-shaped configuration created at the edges of radiators by rectangular slits with length of L_{s2} and width of W_{s2} is added to the Ant. 1, creating MAVA (i.e., Ant. 2). As can be obviously seen from Fig. 3.2b impedance matching is relatively improved. The parameters of the designed antennas are optimised by parametric study using CST Microwave Studio software.

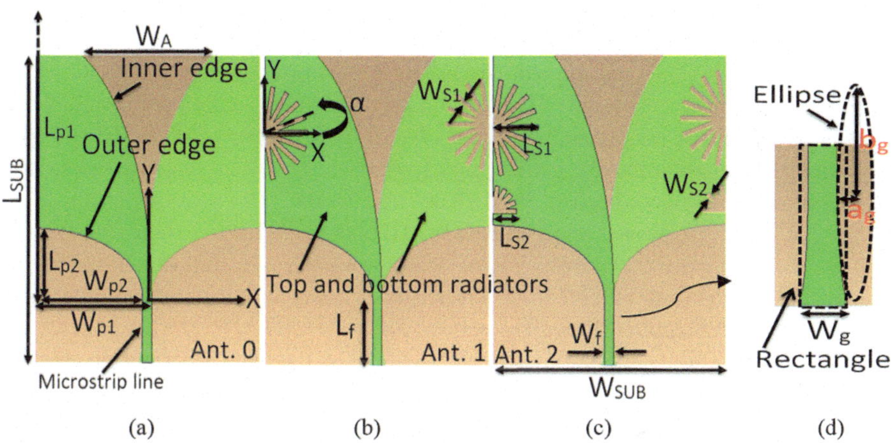

Fig. 3.1 Modification of the designed antennas: (**a**) CAVA (Ant. 0), (**b**) SSCAVA (Ant. 1), (**c**) MAVA (Ant. 2) and (**d**) zoom-in of the ground plane configuration

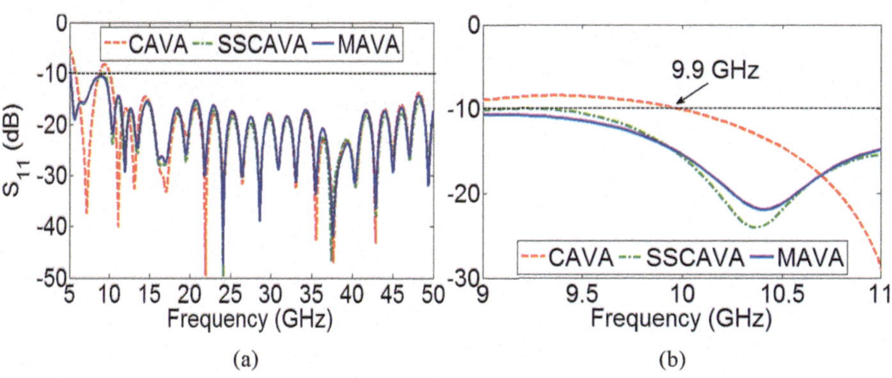

Fig. 3.2 Simulated magnitude of the reflection coefficient of the designed antennas at bandwidth of (**a**) 5–50 GHz and (**b**) 9–11 GHz

To further demonstrate the effectiveness of the sun-shaped elements on extending low end of frequency band, surface current distribution for both the CAVA and MAVA at 9 GHz is shown in Fig. 3.3. As can be clearly seen from Fig. 3.3 the CAVA has intense current in regions 'A' at the edges of the top and bottom radiators that creates radiation in vertical direction (i.e. with respect to *Y*-direction as endfire direction). To mitigate unwanted current in regions 'A', the rectangular slits at sun-shaped configuration are used to alter the surface currents from regions 'A' to 'B' and as a result the antenna radiates along *Y*-direction. On the other hand, the effective electrical length of the current path of the antenna is increased by the slits, resulting in extending the low end of frequency band from 9.9 to 5 GHz.

3.2.2 Dielectric Lens Design

Radiation pattern characteristics of the CAVAs are distorted owing to travelling current in the termination portion of the flares causing instable radiation characteristics (e.g. *E*-plane tilt of beam) and decrease of gain (Constantine 2005). To stabilise radiation pattern and increase gain at high frequencies, various shapes of dielectric lenses or directors have been utilised (Kota and Shafai 2011; Molaei et al. 2014; Bourqui et al. 2010; Nassar and Weller 2015; Teni et al. 2013) to extend the antenna substrate and direct the electromagnetic waves toward the endfire direction. In this study, HEDL was found a best option in terms of being cost effective and easy for fabrication over dielectric directors.

Figure 3.4 exhibits the simulated electrical field intensity distribution on *XY*-plane for the designed antennas at frequencies of 20, 30, 40 and 50 GHz, demonstrating that the endfire radiation status is apparent. It can be clearly viewed from

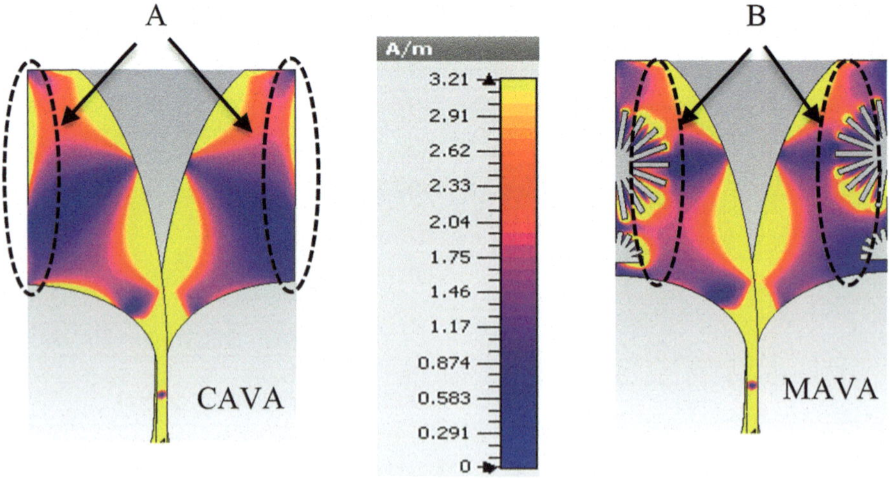

Fig. 3.3 Surface current distribution of the designed CAVA and MAVA at 9 GHz

Fig. 3.4 that the main beam for both CAVA and MAVA is divided into two beams, designated with 'A', as the frequency increases, obviously resulting in decrease of antenna gain. To enhance gain and focusing wave along *Y*-direction, MAVA with HEDL is used and the proposed antenna is referred to as MAVA-HEDL. Two dielectric lenses as extension of antenna substrate have been investigated: one the dielectric lens with major radius of $W_{SUB}/2$ and another dielectric lens with major radius of $W_A/2$ as shown in Fig. 3.5a. Note that the minor radius of the dielectric lens, *R*, is fixed at 15 mm for the two cases. It can be seen from Fig. 3.4 that the MAVA-HEDL case I is not effective since the main beam is still divided at higher frequencies, designated with 'B', while the MAVA-HEDL case II is alleviated of splitting beam at higher frequencies and gain is increased, as expected.

Figure 3.5b shows the gain variation of the two cases at the frequency range of 20–50 GHz. It can be clearly seen from Fig. 3.5b that high gain is achieved in case II at the depicted frequency range and it reaches up to 14 dB GHz at 50 GHz. Here and through the book the maximum gain of the antenna toward the boresight direction is presented.

In order to further demonstrate superiority of the use of case II against case I, the simulated co- and cross-polarisation of the *H*- and *E*-plane radiation patterns at 40 GHz are shown in Fig. 3.6. It is clearly seen from the co-polarisation of *H*- and *E*-planes that the HPBM's case II is narrower than case I. On the other hand, the *E*- and *H*- plane cross-polarisation level <–22 dB in the main beam direction is achieved by MAVA-HEDL (case II).

A parametric study on the various radius values of extension, *R*, is performed for the MAVA-HEDL (case II), as shown in Fig. 3.7. It is clear that the highest gain is attained at $R \geq 15$ mm.

3.3 Results and Discussion

By performing parametric study on different parameters including gain and F-to-B ratio in simulation process, the optimal parameters of the designed MAVA-HEDL (case II) have been achieved as follows: $W_{SUB} = 30$ mm, $L_{SUB} = 40$ mm, $W_A = 18.76$ mm, $W_{p1} = (W_{SUB} + W_f)/2$, $W_{p2} = (W_{SUB} - W_f)/2$, $L_{p1} = 2.165 \times W_{p1}$, $L_{p2} = 8.8$ mm, $\alpha = 20°$, $W_{s1} = 0.6$ mm, $L_{s1} = 5.8$ mm, $W_{s2} = 0.6$ mm, $L_{s2} = 3$ mm, $R = 15$ mm, $W_f = 1.19$ mm, $L_f = 8.5$ mm, $b_g = 8$ mm, $W_g = 2$ mm and $a_g = (W_g - W_f)/2$ mm.

The gain variation of the CAVA, MAVA and MAVA-HEDL (case II) at the entire frequency range is depicted in Fig. 3.8a. It can be seen from Fig. 3.8b that the CAVA gain is low at lower frequencies, e.g. 1.4 dB at 5 GHz and 4 dB at 6 GHz, respectively, whereas the MAVA has improved gain at lower frequencies, i.e. 4.3 dB at 5 GHz and 2 dB at 6 GHz. However, the gain of MAVA decreases after 35 GHz due to longer length of the current path on slits which causes more loss at higher frequencies. As shown in Fig. 3.8c, the antenna gain has been improved at higher frequencies by the HEDL, resulting in its gain reaching to its maximum value of 14 dB at 50 GHz.

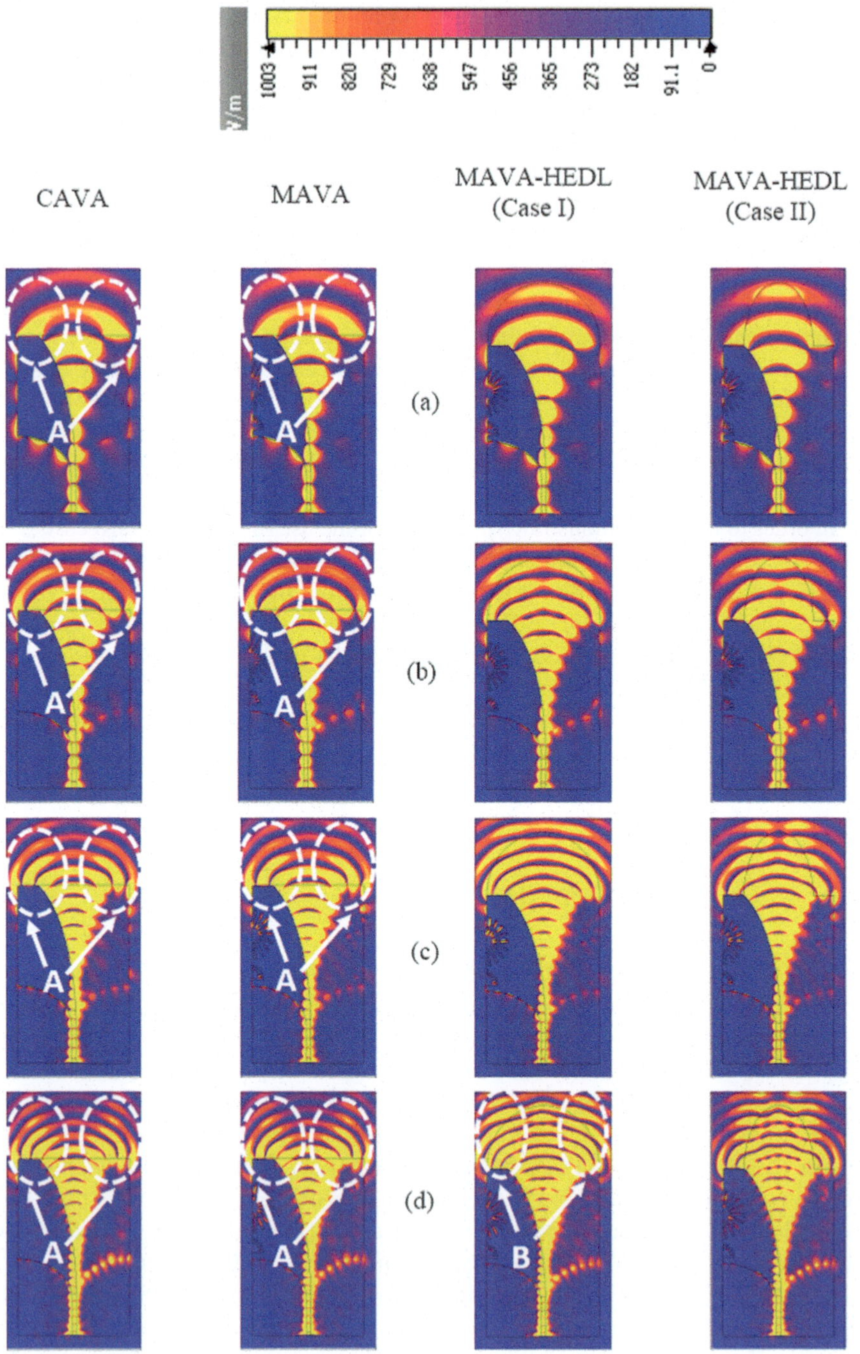

Fig. 3.4 Simulated E-field distribution on *XY*-plane for the designed antennas at (**a**) 20 GHz, (**b**) 30 GHz, (**c**) 40 GHz and (**d**) 50 GHz

To specify the worthiness of the proposed MAVA-HEDL, its operating bandwidth, cross-polarisation levels and gain at highest frequency of bandwidth are compared with the referenced AVAs in accordance with the used size and dielectric constant as shown in Table 3.1. As shown, the MAVA-HEDL exhibits high gain at the highest frequency of operating bandwidth (i.e. 14 dB at 50 GHz) and cross-

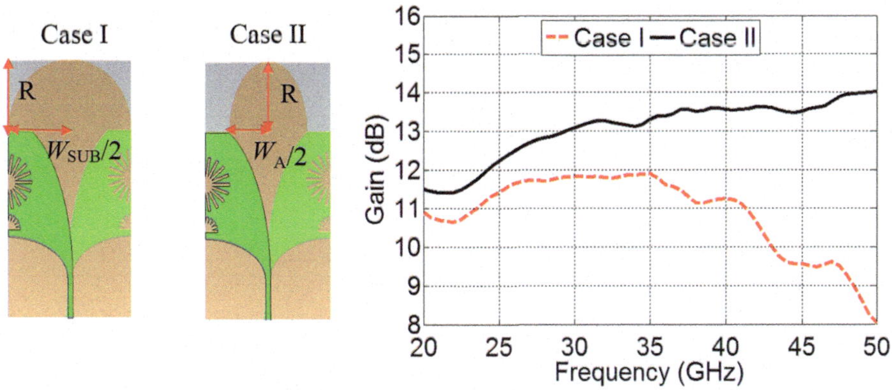

(a) (b)

Fig. 3.5 (**a**) Configuration of the case I and case II of the proposed MAVA-HEDL and (**b**) gain variation for case I and case II at higher frequencies (20–50 GHz)

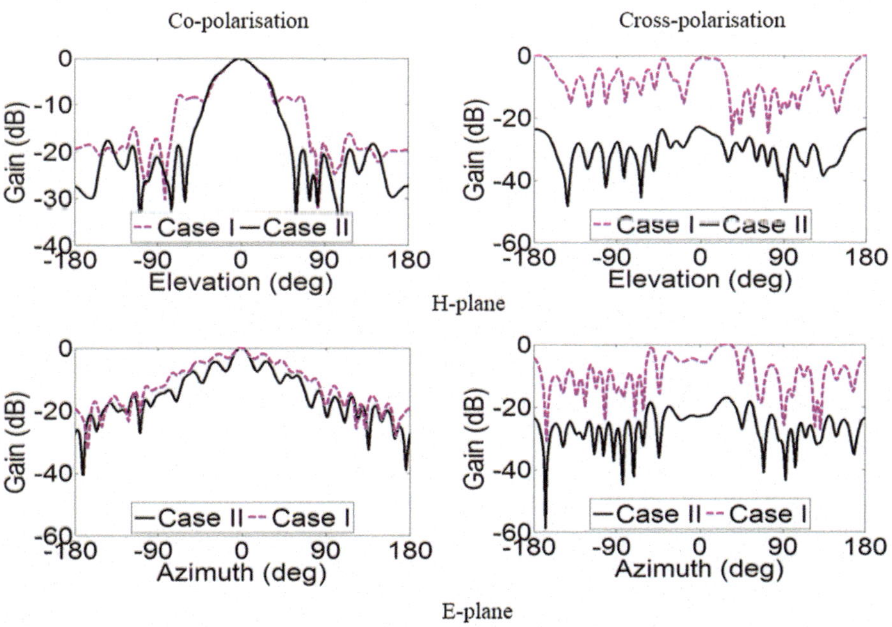

Fig. 3.6 The simulated co- (left) and cross-polarisation (right) of the *H*- and *E*-plane radiation patterns for the case I and case II of the MAVA-HEDL at 40 GHz

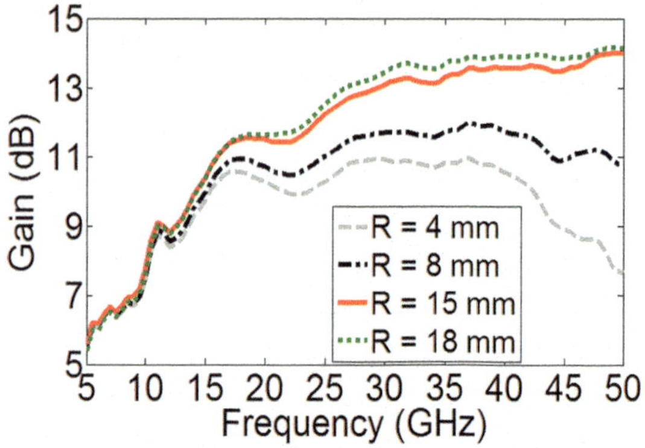

Fig. 3.7 Variation of the gain with frequency for the various radius values of R for the MAVA-HEDL (case II)

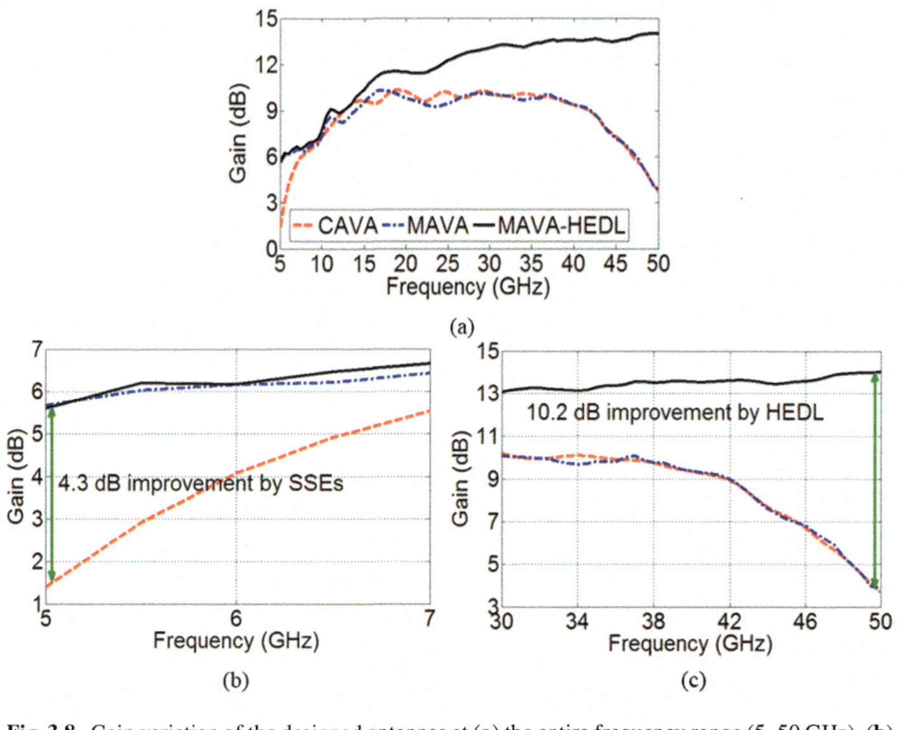

Fig. 3.8 Gain variation of the designed antennas at (**a**) the entire frequency range (5–50 GHz), (**b**) lower frequencies (5–7 GHz) and (**c**) higher frequencies (30–50 GHz)

polarisation level <−22 dB over the whole operating frequency band. It should be mentioned that the cross-polarisation level <−25 dB is obtained by balanced AVA (BAVA) described in Molaei et al. (2014) since BAVAs are able to improve cross-polarisation level of AVAs.

The simulated radiation efficiency from 5 to 50 GHz is shown in Fig. 3.9a. As shown, the radiation efficiency is from 55 to 98% for the designed antennas. The radiation efficiencies of both MAVA and MAVA-HEDL vary from 60 to 80% while

Table 3.1 Performances of the referenced antipodal Vivaldi antennas and the proposed MAVA-HEDL

References	Dimensions (mm^3)	ε_r	O.B.W. (GHz)	Gain at high freq.	Cross-pol. level (dB)
Li et al. (2016)	70 × 166 × 0.76	3.48	4.7–20	14.2	< −10
Chamaani et al. (2011)	94 × 86.2 × 0.76	3.5	2.6–11	9 dB	< −11.5
Daniel et al. (2011)	50 × 150 × 0.4	2.17	4–40	10 dB	N.M.
Walden (2013)	50 × 80 × 0.508	2.2	7–50	11 dB	N.M.
Karahan and Armagan Sahinkaya (2014)	27 × 63.5 × 0.813	3.55	6–18	11.8 dB	N.M.
Chuang et al. (2016)	17.77 × 30.48 × 0.8	3.55	9.24–18.14	N.M.	N.M.
Hshieh et al. (2013)	28.7 × 46 × 0.305	3.38	5.5–40	N.M.	N.M.
Fei et al. (2011)	48 × 60 × 1	4.6	2.4–14	7.5 dB	N.M.
Bai et al. (2011)	64 × 61 × 0.254	2.3	4–50	12 dB	N.M.
Natarajan et al. (2015)	42 × 36 × 1.6	4.4	3.7–18	4.5 dB	< −20
De Oliveira et al. (2015)	36.30 × 59.8 × 0.64	6.15	5.6–16	8 dB	N.M.
Dastranj (2015)	31.2 × 45 × 0.813	3.55	6–18	8.5 dB	< −20
Puskely et al. (2016)	8 × 59 × 0.5	2.33	25–42	10.2	< −10
Teni et al. (2013)	50 × 66.4 × 1	4.5	4–30	7 dB	N.M.
Kota and Shafai (2011)	56 × 106 × 1.5	2.5	6–20	11 dB	N.M.
Molaei et al. (2014)	50 × 166 × 3.15	2.55	3–18	12 dB	< −25
Bourqui et al. (2010)	44 × 80 × 9.2	2.94	2.4–18	N.M.	N.M.
Nassar and Weller (2015)	66 × 140 × 1.5	2.94	2–32	−1 dB	N.M.
Kim and Chang (2004)	N.M.	2.2	8–50	10	< −12
Moosazadeh and Kharkovsky (2016)	40 × 90 × 0.508	3.38	3.4–40	14 dB	N.M.
This work, MAVA-HEDL	30 × 55 × 0.528 (with lens)	3.38	5–50	14 dB	< −22

'O.B.W.' and 'N.M.' state the operating bandwidth and not mentioned, respectively

it varies from 55 to 75% for CAVA at lower frequencies from 5 to 6 GHz as shown in Fig. 3.9b. There is a drop in the simulated radiation efficiency below 8 GHz; it can be due to losses imposed by slits. The radiation efficiency decreases with frequency with all designs as shown in Fig. 3.9a. The reason is that the radiation efficiency of the antennas depends on thickness and material of substrate used in antennas which causes losses at higher frequencies. Also, there is jitter behaviour in the radiation efficiency at frequency range from 8 to 12 GHz, especially for the MAVA-HEDL, yet the gain is not affected as shown in Fig. 3.8a. The reason is that when the MAVA-HEDL operates at lower frequencies, the current on the antenna may be in-phase (i.e. waves add together in far-field region) and focused in the same direction, thereby causing a higher gain.

The simulated co- and cross-polarisation of *E*-plane (*XY*-plane) and *H*-plane (*YZ*-plane) radiation patterns of the designed CAVA and MAVA-HEDL at the different frequencies are shown in Fig. 3.10, indicating that the all designed antennas have endfire direction feature with the main lobe in *Y*-direction (Fig. 3.1). It can be

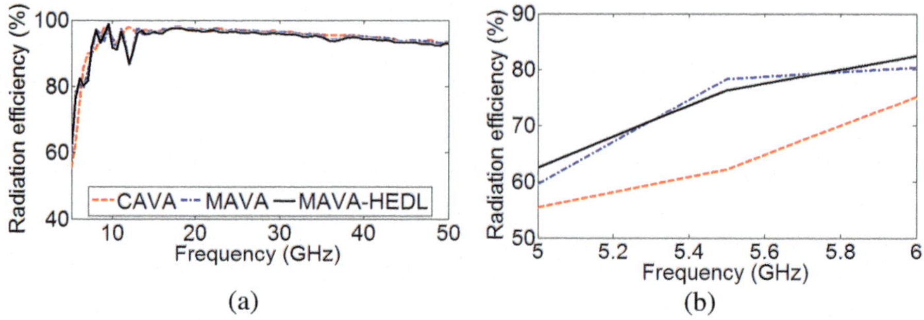

Fig. 3.9 Simulated radiation efficiency of the designed antennas at (**a**) the entire frequency band (5–50 GHz) and (**b**) lower frequencies (5–6 GHz)

clearly seen that the HPBW of the antennas is wide at lower frequencies and exhibits less directivity, while as the frequency increases the patterns show higher directivity. According to the depicted radiation patterns, smaller side lobe level can be seen for the MAVA-HEDL compared with the CAVA. In addition, the cross-polarisation level <−22 dB is achieved for the MAVA-HEDL in the entire frequency band.

Effectiveness of the used sun-shaped elements and dielectric lens on improvement of the F-to-B ratio is illustrated in Fig. 3.11. As shown, a significant improvement of the F-to-B ratio for the MAVA-HEDL is achieved compared to the CAVA. The maximum F-to-B ratio (26.8 dB) is attained with the MAVA-HEDL at a frequency of 23 GHz which is approximately twice of that for the CAVA (i.e. 14.2 dB).

According to the *E*-plane radiation patterns depicted in Fig. 3.10b, an investigation on HPBW and tilt of beam is carried out and obtained results are summarised in Table 3.2. It can be clearly found that the MAVA-HEDL has narrower *E*-plane HPBW compared with the CAVA and MAVA. For instance, the HPBW at 50 GHz for the CAVA and MAVA are 48.1° and 48.6°, respectively, while it is 12.2° for the MAVA-HEDL.

In most Vivaldi antennas, the main lobe for *E*-plane is tilted from its nominated axial direction at higher frequencies. Investigation on the *E*-plane tilt of beam for the designed antennas for some frequencies is also performed and shown in Table 3.2. It can be obviously seen that the tilt of beam is significantly reduced for the MAVA-HEDL than that for both the CAVA and MAVA, especially at higher frequencies. For instance, the *E*-plane tilt of beam is 21° at 44 GHz and −16° at 50 GHz for the CAVA while it is 0° at 44 GHz and −1° at 50 GHz for the MAVA-HEDL. Compared with some of the referenced AVAs, it is worthwhile to point that the *E*-plane tilt of beam at 15 GHz in (Kota and Shafai 2011; Molaei et al. 2014) is 4° with the used dielectric lenses, while it is 0° for the designed MAVA (without dielectric lens).

The time domain behaviour of the proposed antenna was investigated as well. First, in order to estimate time signal distortion, group time delay parameter of the MAVA-HEDL antenna was studied in two angles, namely $\theta = 0°$ (i.e. *Y*-direction in

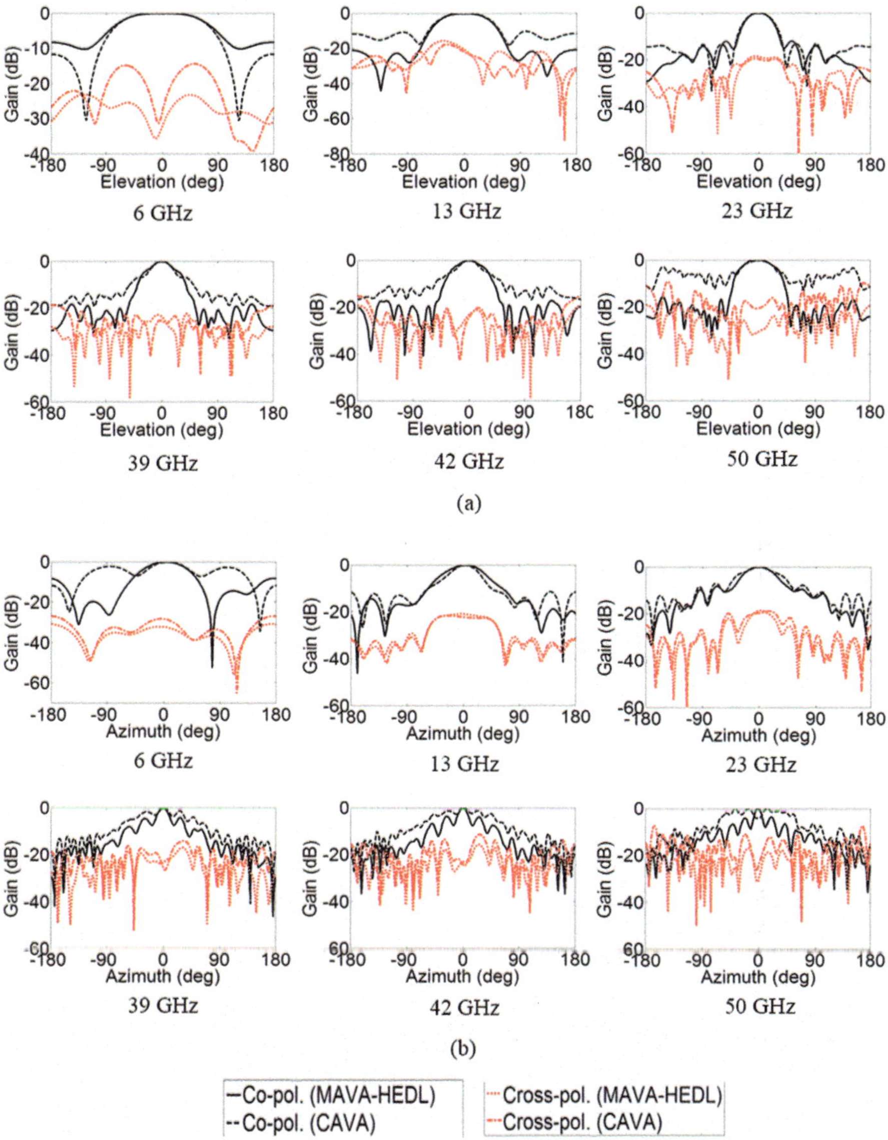

Fig. 3.10 The simulated co- and cross-polarisation of (**a**) *H*-plane and (**b**) *E*-plane radiation patterns at different frequencies for the CAVA and MAVA-HEDL

Fig. 3.1) and $\theta = 30°$ as shown in Fig. 3.12a. To do this, two identical antennas were placed face to face with distance of 100 mm to fulfil transmitting and receiving signals. It can be clearly seen from Fig. 3.12b that the supreme group time delay of the proposed antenna is around 1.3 ns with fluctuation of less than 0.1 ns in most frequencies. In other words, this antenna is able to transmit and receive signals without serious distortion.

Figure 3.13 shows the fidelity factor which is specified by the absolute value of the maximum of the cross-correlation coefficient, i.e. level of similarity between transmitted and received signals (Quintero et al. 2011). As shown, the fidelity value is attained over 0.97 for case of $\theta = 0°$; it means that the designed antenna operates a nearly distortionless transmitted signal.

The magnitude of transmission coefficient, S_{21} in dB, has also been assessed for the designed antennas. Figure 3.14 exhibits that the HEDL increases the magnitude of transmission coefficient between 5–8 GHz and 12.5–30 GHz. This fact also confirms higher directivity of the proposed antenna compared to the CAVA and MAVA.

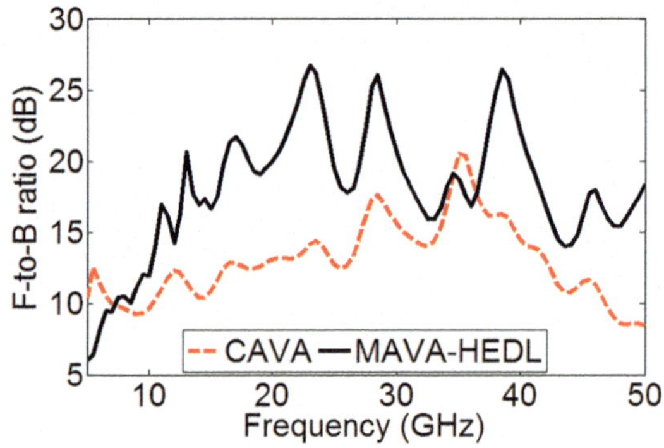

Fig. 3.11 Variation of the F-to-B ratio with frequency for the designed CAVA and MAVA-HEDL

Table 3.2 *E*-plane HPBW and tilt of beam for the designed antennas

Freq. (GHz)	*E*-plane HPBW, degree			*E*-plane tilt of beam, degree		
	CAVA	MAVA	MAVA-HEDL	CAVA	MAVA	MAVA-HEDL
6	58.8	62.2	63.9	7	3	4
11	52.1	50.7	49.6	5	4	3
15	43.6	45	44.6	2	0	0
16	46.5	39.4	39.4	3	0	0
22	60.7	65.2	48.8	1	1	1
27	54.4	53.5	36.4	2	−1	−1
32	50.5	50.4	29.9	1	3	0
37	51.2	27.4	21.5	1	1	0
44	97.1	96.9	15.8	21	21	0
45	95.2	83	15.4	−20	−22	0
46	93.7	79.9	14.7	−19	−21	0
47	97.1	84	13.7	−17	−20	0
48	118.9	118.4	13.1	−16	−18	−1
49	51.5	50.9	12.7	−16	−17	−1
50	48.1	48.6	12.2	−16	−17	−1

3.4 Microwave and Millimetre Wave Imaging Applications of the MAVA-HEDL

3.4.1 Fabricated Antenna and Measurement Imaging Approach

A prototype of the MAVA-HEDL (case II) with the optimal parameters is fabricated and tested with Agilent N5225A Performance Network Analyser (PNA) to validate the simulated result of S_{11}. A photo of the fabricated antenna with 2.4 mm connector is shown in Fig. 3.15a. The measured and simulated magnitude of the reflection coefficient of the proposed antenna is plotted in Fig. 3.15b, indicating that the antenna has bandwidth from 5 to 50 GHz for S_{11} <−10 dB. As shown in Fig. 3.15b, there is good agreement between measured and simulated results except some discrepancies for measured result which can be due to inaccuracy in fabrication of antenna, irregularity of substrate drilling to create holes for screwing connector and calibration error of PNA at higher frequencies.

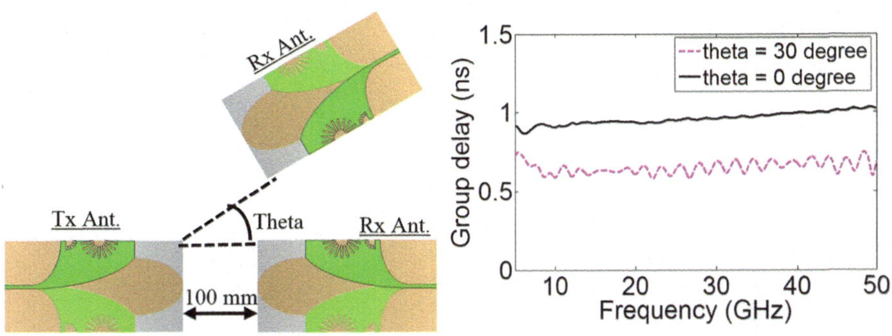

(a) (b)

Fig. 3.12 (**a**) Two antennas arrangement at $\theta = 0$ and 30° and (**b**) group delay of the proposed MAVA-HEDL

Fig. 3.13 Cross-correlation/fidelity factor for the transmitted and received signals of the proposed MAVA-HEDL for $\theta = 0$ and 30°

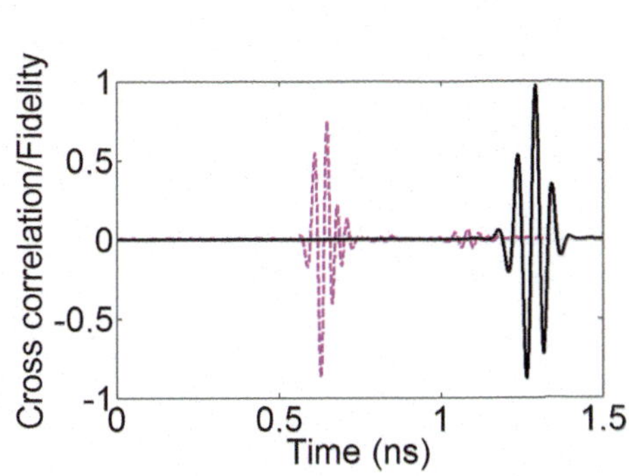

To demonstrate capability of the proposed antenna for high-range resolution imaging of samples made of construction materials, a multifunctional imaging system was used. The complex microwave reflection coefficient over a plane for each sample was measured using MAVA-HEDL, a custom-made imaging system with a commercial PNA and a scanning mechanism as shown in Fig. 3.16 (Kharkovsky et al. 2014). The antenna was moved by the scanning mechanism at standoff distance (i.e. the distance between the antenna and the sample under test), *d*, to obtain raster scans at a scan plane, as shown in Fig. 3.17. The sample was irradiated by the antenna using continuous wave (CW) approach at 101 frequency points and scanned with a 3 mm step size (along *X*- and *Y*-axes). The complex reflection coefficient was measured over a scan area and saved as a binary file. The three- and two-dimensional (3D and 2D) images of the sample were generated using the file and a synthetic aperture radar (SAR) algorithm (Case et al. 2011; Cumming and Wong 2005).

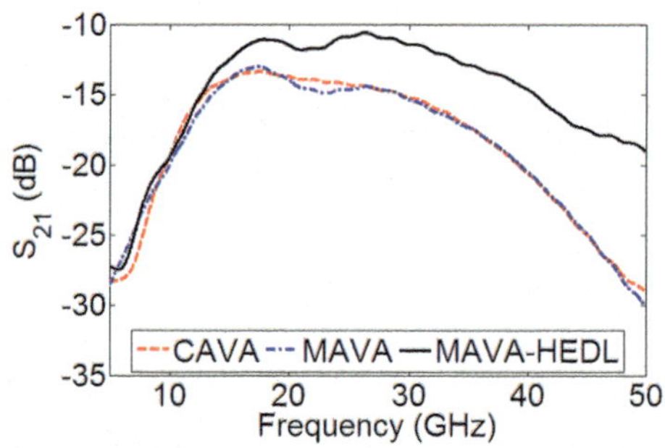

Fig. 3.14 The magnitude of transmission coefficient, S_{21} in dB, for the designed antennas

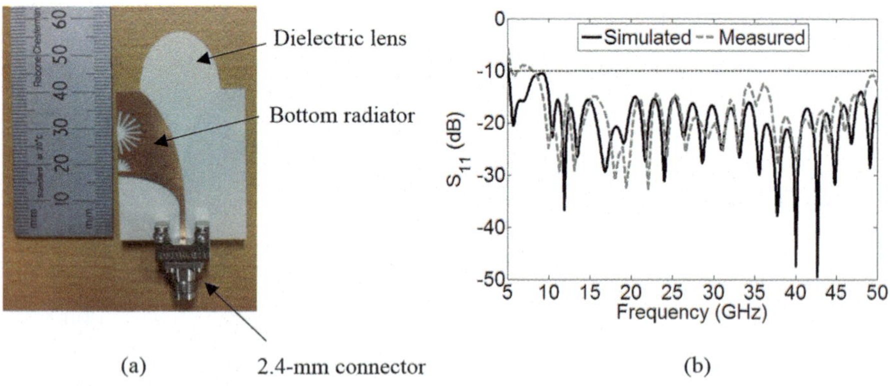

Fig. 3.15 (**a**) Photograph of the fabricated MAVA-HEDL (case II) with 2.4 mm connector and (**b**) simulated and measured magnitude of the reflection coefficient versus frequency of the proposed antenna

3.4.2 *Imaging of Rubber Discs in Construction Foam Layered Structure*

A few samples were prepared and imaged. One of the samples was a construction foam layered structure with dimensions of 240 mm by 240 mm by 90 mm including six equally thick layers (referred to as Sample 1 here). Four rubber discs (each of diameter of 25 mm and thickness of 2 mm) in two groups of two were placed at interfaces between layers at the depth of 14 mm and 74 mm, respectively, as shown in Fig. 3.18a–c. A perspective view of the 3D image of Sample 1, obtained over a scan area of 210 mm by 210 mm and $d = 10$ mm, is shown in Fig. 3.18d. Figure 3.18c, d shows 2D images as horizontal slices of the 3D image when high-frequency signal was focused at the plane of location of upper discs and lower discs, respectively. Clear and sharp indications of all four discs at uniform background can be seen in the 3D image (c.f. Fig. 3.18d). Indications of only two discs located at the same depth are visible in each 2D image (c.f. Fig. 3.18c–d). It should be noted that indications of the discs at different depths are very similar. Overall, the obtained images of Sample 1 clearly demonstrate the capability of the proposed antenna

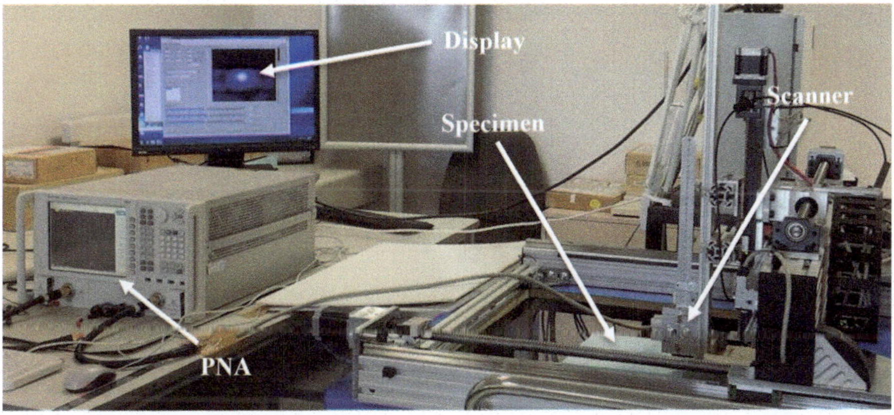

Fig. 3.16 Microwave imaging system in the CIE, Western Sydney University

Fig. 3.17 Schematic of the set-up testing a sample with the proposed antenna (MAVA-HEDL) using a raster scan

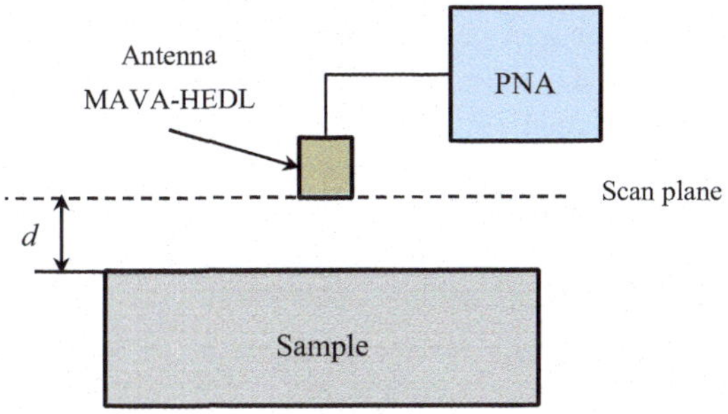

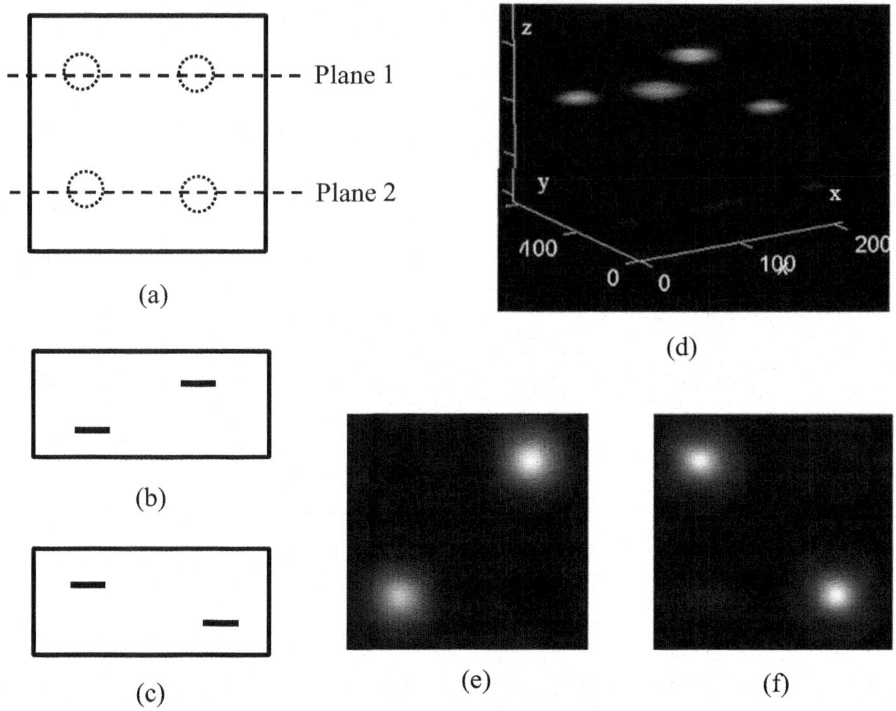

Fig. 3.18 Sample 1: (**a**) schematic of top view with indication of embedded discs, (**b**) cross-sectional side view at Plane 1 and (**c**) cross-sectional side view at Plane 2, and (**d**) 3D image of the discs and its horizontal slices at the depth of (**e**) 14 mm and (**f**) 74 mm

incorporated in the UWB imaging technique for the detection and evaluation of targets such as rubber discs embedded in construction foam at different depths.

3.4.3 Metal Rods Embedded in Layered Plasterboard Sheet Structure

Other two samples included 12 mm diameter metal rods embedded in plasterboard sheet—air gap layered structure located on the top of a 250 mm cubic concrete block as shown in Figs. 3.19a and 3.24a. Each rod has the length of 250 mm and one rod was located in the air gap between plasterboard sheets 1 and 2 while another rod was located in the air gap between plasterboard sheets 2 and 3. Plasterboard sheets 1 and 3 had thickness of 14 mm and the thickness (referred to as t_2) and material of sheet 2 were varied. In Sample 2 (Sample 3) the two rods were non-aligned (aligned) with separation distance of $L = 60$ mm (0 mm) as shown in Figs. 3.19a, b and 3.24a, b, respectively.

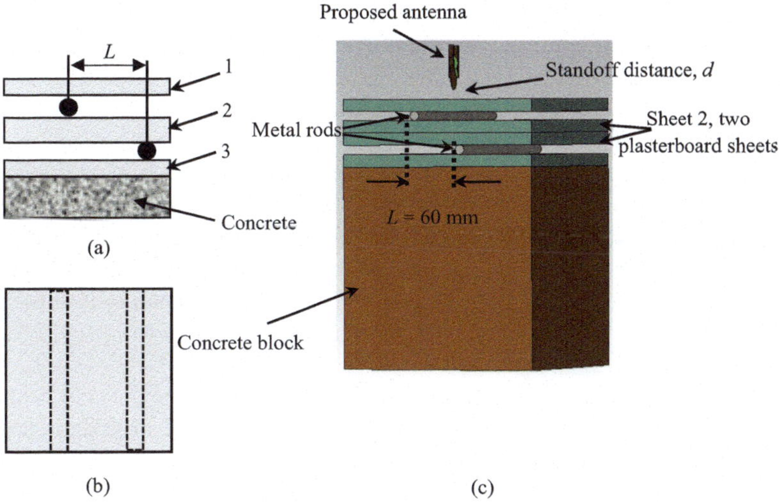

Fig. 3.19 Sample 2: (**a**) cross-sectional side view, (**b**) top view and (**c**) a model of Sample 2 for non-aligned metal rods ($L = 60$ mm) and sheet 2 with two plasterboard sheets in CST

- **Simulation results for Sample 2**

In order to investigate the behaviour of the magnitude and phase of the reflection coefficient and the electric field intensity distribution, models of the proposed antenna and concrete block (dielectric constant of 4.5 and loss tangent of 0.0125) along with plasterboard sheets (dielectric constant of 3.01 and loss tangent of 0.036) possessing non-aligned metal rods were created and investigated in CST. One model consists of the proposed antenna irradiating Sample 2 and is shown in Fig. 3.19c. The model without metal rods was also simulated and used as a reference.

Figure 3.20a, b shows the magnitude and phase of the reflection coefficient at a frequency range of 6–30 GHz when the antenna was located at a standoff distance of 12 mm of Sample 2 above the upper metal rod. It can be seen from Fig. 3.20 that there is a difference between values of the magnitude and phase of the reflection coefficient for Sample 2 without and with non-aligned metal rods over the frequency range from 6 to 30 GHz. It can be expected that this difference will provide indication of the rods in images of the sample.

Electrical field intensity distribution for the proposed antenna and Sample 2 without and with non-aligned metal rods at frequencies of 12.29 GHz and 16.22 GHz is depicted with colour bar in Fig. 3.21. Figure 3.21 shows reflecting and scattering electromagnetic wave when antenna is irradiating upper metal rod in the layered structures of the plasterboard sheets and as a result it significantly changes the electric field distribution at the interface between free space and upper metal rod. The following observations can be made from Fig. 3.21. The reflection from air-sheet interference occurs at different frequencies (Fig. 3.21a, b). In addition, the comparison of field distributions at the two frequencies clearly indicates different values of depth of penetration in concrete block at different frequencies; namely, penetration depth at relatively higher frequency (i.e. at 16.22 GHz) is less than that at lower frequency (i.e. at 12.29 GHz). Therefore, it could be expected that the signals at higher frequencies (>16.22 GHz) would not sense the bottom side of the concrete block. This observation explains the behaviour of the magnitude and phase of reflection coefficient at different frequencies as shown in Fig. 3.20.

Figure 3.22a, b shows the magnitude and phase of the reflection coefficient at a frequency range of 6–30 GHz when the antenna was located above the lower metal rod at a standoff distance of 12 mm of Sample 2. It can be seen from Fig. 3.22 that there are some differences between values of the magnitude and phase of the reflection coefficient for Sample 2 without and with non-aligned metal rods at the lower and middle frequencies. It can be expected that this difference will provide indication of the lower rods in images of the sample.

Electrical field intensity distribution for the proposed antenna and Sample 2 without and with non-aligned metal rods at a frequency of 12.77 GHz is depicted with colour bar in Fig. 3.23 when antenna was located along the lower metal rod. Figure 3.23 shows that the electromagnetic waves are scattered at the interface between free space and sheets. It can be clearly seen from Fig. 3.23 (right) that the electromagnetic waves are reflected and change the electric field distribution when antenna is irradiating lower metal rod in the layered structures of the plasterboard sheets and significantly resulting in splitting of the main lobe along their propagation.

- **Simulation results for Sample 3**

In order to investigate the behaviour of the magnitude and phase of the reflection coefficient and the electric field intensity distribution, models of the proposed antenna and concrete block along with plasterboard sheets possessing aligned metal rods were created and investigated in CST. One model consists of the proposed antenna irradiating the upper metal rod of Sample 3 and is considered as shown in Fig. 3.24. The case when there are no metal rods was also simulated and used as a reference.

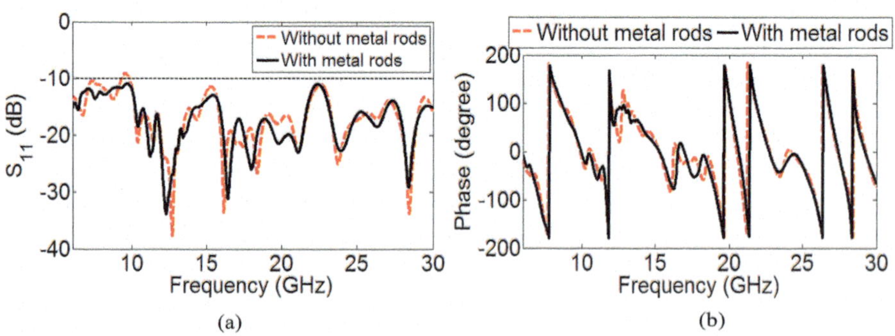

Fig. 3.20 The simulated (**a**) magnitude and (**b**) phase of the reflection coefficient for Sample 2 when antenna was located above the upper metal rod

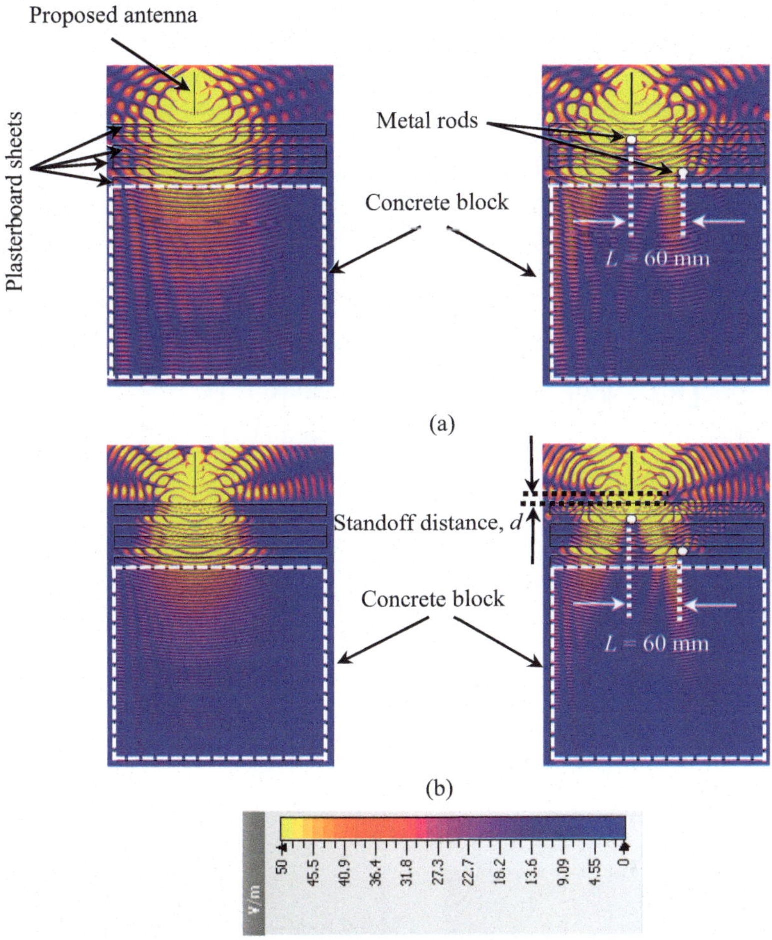

Fig. 3.21 Electrical field intensity distribution (*YZ*-plane) at standoff distance of 12 mm of Sample 2 above the upper metal rod indicated in Fig. 3.19c without (left) and with non-aligned ($L = 60$ mm) metal rods (right) embedded in the plasterboard sheets at the frequencies of (**a**) 12.29 GHz and (**b**) 16.22 GHz

Figure 3.25a, b shows the magnitude and phase of the reflection coefficient at a frequency range of 6–30 GHz when the antenna was located at a standoff distance of 12 mm of Sample 3. It should be mentioned that metal rods are aligned ($L = 0$ mm) with the proposed antenna. It can be seen from Fig. 3.25 that there is a difference between values of the magnitude and phase of the reflection coefficient for Sample 3 without and with aligned metal rods over the entire frequency range from 6 to 30 GHz.

Electrical field intensity distribution for the proposed antenna and Sample 3 without and with aligned metal rods embedded in the plasterboard sheets at frequencies of 6.74 and 12.74 GHz is depicted with colour bar in Fig. 3.26. Figure 3.26 demonstrates that the presence of aligned metal rods in the layered structures of the plasterboard sheets significantly changes the electric field distribution at the interface between free space and Sample 3. The following observation can be made from Fig. 3.26. The presence of the metal rods is caused to be not only reflecting electromagnetic wave at the interface between the free space and Sample 3 but also splitting electromagnetic waves. This observation explains the behaviour of the magnitude and phase of reflection coefficient at different frequencies as shown in Fig. 3.25.

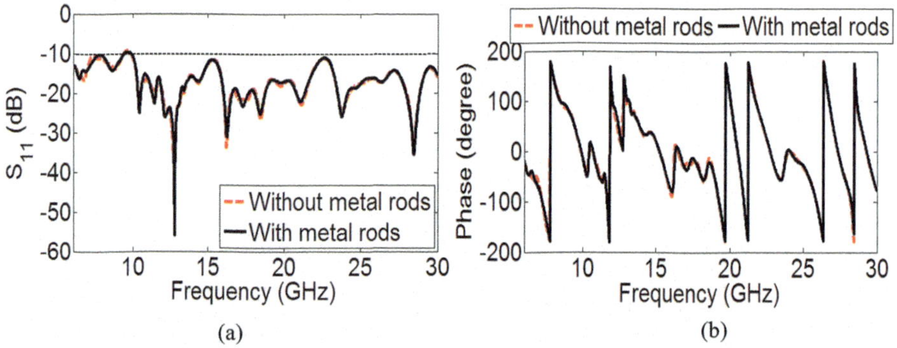

Fig. 3.22 The simulated (**a**) magnitude and (**b**) phase of the reflection coefficient for Sample 2 when antenna was located above the lower metal rod

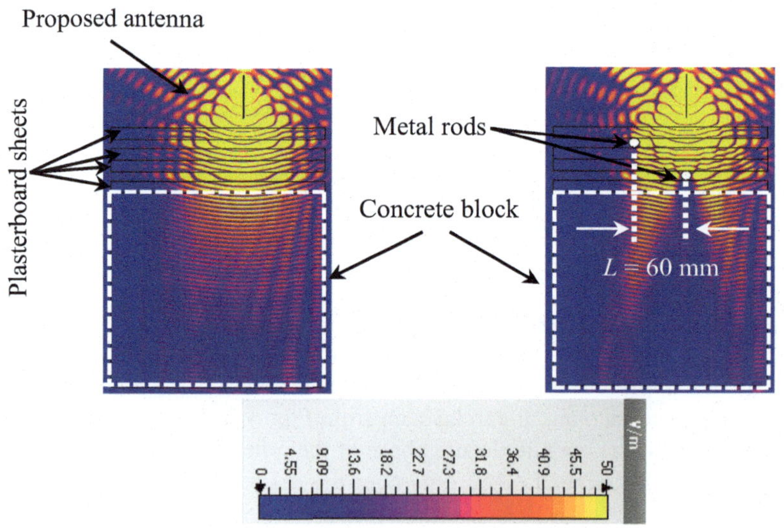

Fig. 3.23 Electrical field intensity distribution (*YZ*-plane) at a standoff distance of 12 mm of Sample 2 above the lower metal rod without (left) and with non-aligned (L = 60 mm) metal rods (right) embedded in the plasterboard sheets at a frequency of 12.77 GHz

- **Imaging results for Sample 2**

Figure 3.27a–d shows 2D images (scan area of 150 mm by 150 mm, $d = 12$ mm) of Sample 2 at different values of thickness of layer 2. It can be seen that a separate indication of rods located at different depths is clearly visible at t_2 from 28 to 8 mm. Indications of the rods in this case could not be seen separately at $t_2 = 4$ mm.

- **Imaging results for Sample 3**

Distinguishing indications of the rods in Sample 3 is a challenging task since they are located one under another and indication of the lower rod can be masked by indication of the upper rod and/or the indications of both rods can be disturbed due

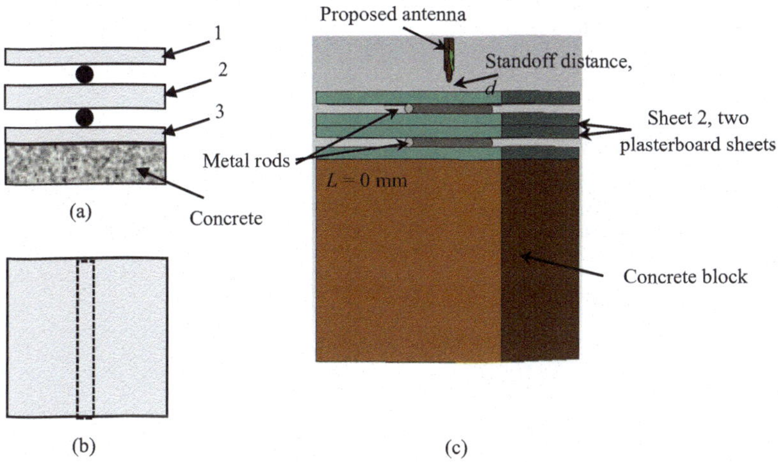

Fig. 3.24 Sample 3: (**a**) cross-sectional side view, (**b**) top view and (**c**) model of Sample 3 for aligned metal rods ($L = 0$ mm) and sheet 2 with two plasterboard sheets in CST

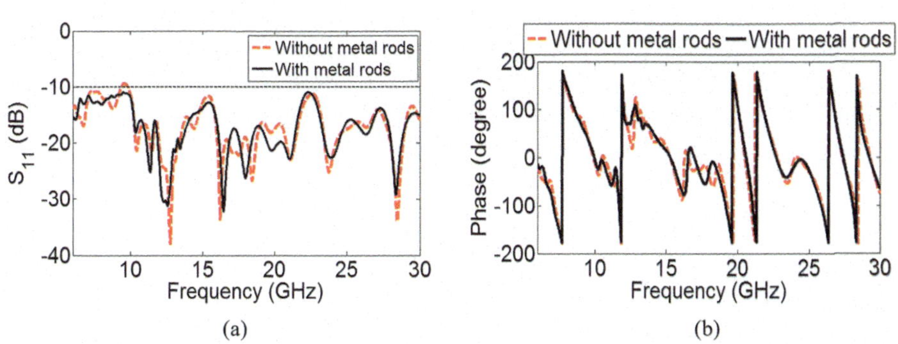

Fig. 3.25 The simulated (**a**) magnitude and (**b**) phase of the reflection coefficient for the CST model indicated in Fig. 3.23c

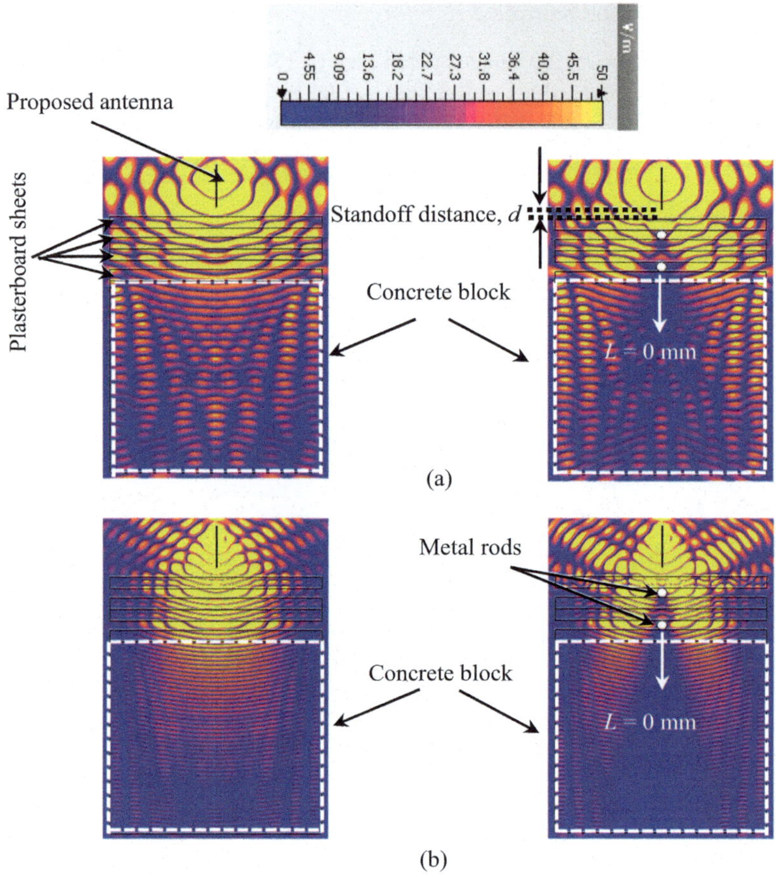

Fig. 3.26 Electrical field intensity distribution (*YZ*-plane) for Sample 3 indicated in Fig. 3.23c without (left) and with aligned ($L = 0$ mm) metal rods (right) embedded in the plasterboard sheets at the frequencies of (**a**) 6.74 GHz and (**b**) 12.74 GHz

to multiple reflections of high-frequency signals between rods. The investigation of Sample 3 showed that the indications of the upper rods were clearly visible in 2D images of Sample 3 as shown in Fig. 3.28a–c. As expected, the indication of the lower rod was disturbed; however, it was visible at some thickness of layer 2 as shown in Fig. 3.28a–c for t_2 from 14 to 42 mm.

3.5 Summary

A small and modified antipodal Vivaldi antenna with size of 30 × 55 × 0.508 mm^3 (i.e. $0.5\lambda \times 0.9\lambda \times 0.005\lambda$, λ is the free-space wavelength corresponding to the lowest operating frequency of 5 GHz) and bandwidth from 5 to 50 GHz was developed

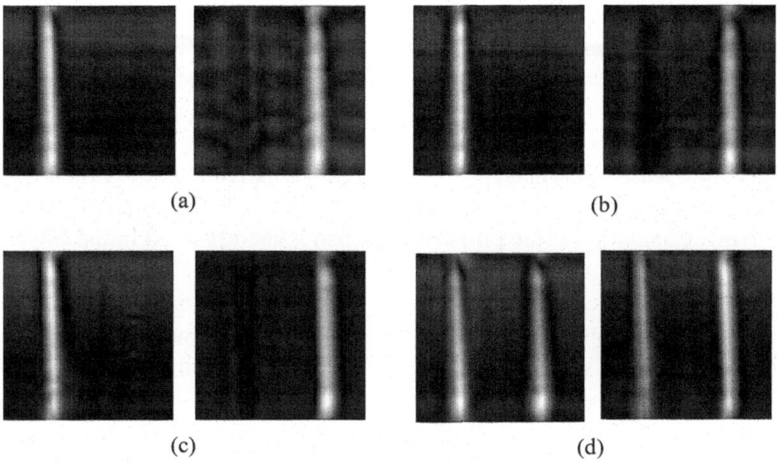

Fig. 3.27 2D images (horizontal slices) at thickness of layer 2 of (**a**) 28 mm (two plasterboard sheets), (**b**) 14 mm (one plasterboard sheet), (**c**) 8 mm (foam sheet) and (**d**) 4 mm (paper sheets) when a high-frequency signal was focused on (left) upper rod and (right) lower rod ($L = 60$ mm)

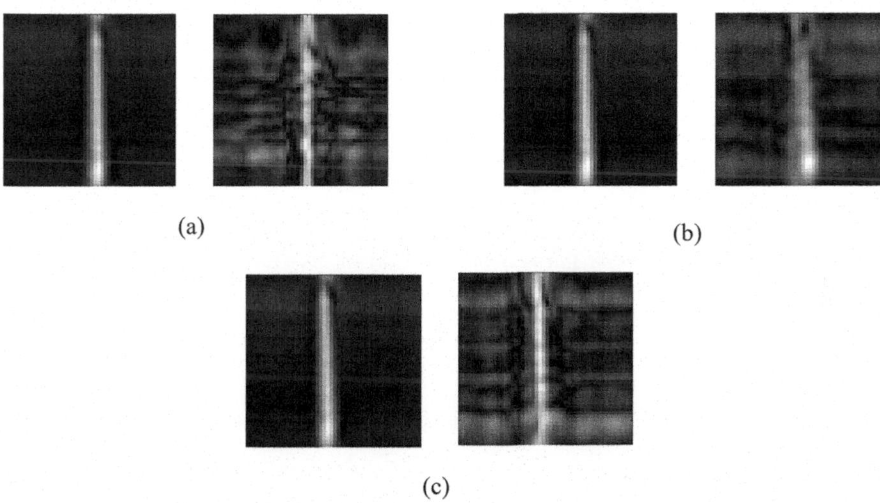

Fig. 3.28 2D images (horizontal slices) at thickness of layer 2 of (**a**) 42 mm (three plasterboard sheets), (**b**) 28 mm (two plasterboard sheets) and (**c**) 14 mm (one plasterboard sheet) when a high-frequency signal was focused on (left) upper rod and (right) lower rod ($L = 0$)

for microwave and millimetre wave imaging of low-loss construction materials and structures. Compare to a similar conventional AVA, lower end of frequency band limitation for $S_{11} < -10$ dB has been extended and gain was increased up to 5.7 dB at lower frequencies by applying rectangular slits at a sun-shaped configuration. A half elliptical shaped dielectric lens as extension of antenna substrate was employed to enhance antenna gain at higher frequencies, and as a result the antenna gain is 14 dB at 50 GHz. In addition, the proposed antenna demonstrates reduced *E*-plane tilt of beam, e.g. $-1°$ at 50 GHz, and lower side lobe level, cross-polarisation level (<-22 dB), *E*-plane half-power beamwidth (e.g. 12.2° at 50 GHz) and improved F-to-B ratio. It was shown that gain of the proposed antenna is >5.7 dB, >10 dB and >13 dB at a frequency range from 5 to 50 GHz, 10 to 50 GHz and 30 to 50 GHz, respectively. Capability of the proposed antenna is incorporated in the UWB sensing technique for the high-range-resolution detection and evaluation of targets such as rubber discs and metal rods embedded in a construction foam structure and in a layered plasterboard structure, respectively. The imaging results show that the proposed antenna can be used for high-range-resolution imaging of a variety of low-loss construction materials as well as other materials.

References

Ahmed, S. S., Schiessl, A., Gumbmann, F., Tiebout, M., Methfessel, S., & Schmidt, L. (2012). Advanced microwave imaging. *IEEE Microwave Magazine, 13*, 26–43.

Bai, J., Shi, S., & Prather, D. W. (2011). Modified compact antipodal Vivaldi antenna for 4–50-GHz UWB application. *IEEE Transactions on Microwave Theory and Techniques, 59*, 1051–1057.

Bourqui, J., Okoniewski, M., & Fear, E. C. (2010). Balanced antipodal Vivaldi antenna with dielectric director for near-field microwave imaging. *IEEE Transactions on Antennas and Propagation, 58*, 2318–2326.

Case, J. T., Ghasr, M. T., & Zoughi, R. (2011). Optimum two-dimensional uniform spatial sampling for microwave SAR-based NDE imaging systems. *IEEE Transactions on Instrumentation and Measurement, 60*, 3806–3815.

Chamaani, S., Mirtaheri, S. A., & Abrishamian, M. S. (2011). Improvement of time and frequency domain performance of antipodal Vivaldi antenna using multi-objective particle swarm optimization. *IEEE Transactions on Antennas and Propagation, 59*, 1738–1742.

Chuang, J.-K., Cheng, Y.-H., & Wang, C.-L. (2016). *Compact and broadband microstrip-fed antenna using antisymmetric tapered probe with triangular element*. In 2016 IEEE International Symposium on Radio Frequency Integration Technology (RFIT). IEEE, pp. 1–3.

Constantine, A. B. (2005). *Antenna theory: Analysis and design. Microstrip antennas* (3rd ed.). New York: Wiley.

Cumming, I. G., & Wong, F. H. (2005). Digital processing of synthetic aperture radar data. *Artech House, 1*, 3.

Daniel, C., Lilly, J., & Auckland, D. (2011). *A compact, cost-effective 4–40 GHz antenna*. In 2011 IEEE International Symposium on Antennas and Propagation (APSURSI).

Dastranj, A. (2015). Wideband antipodal Vivaldi antenna with enhanced radiation parameters. *IET Microwaves, Antennas and Propagation, 9*, 1755–1760.

De Oliveira, A. M., Perotoni, M. B., Kofuji, S. T., & Justo, J. F. (2015). A palm tree antipodal Vivaldi antenna with exponential slot edge for improved radiation pattern. *IEEE Antennas and Wireless Propagation Letters, 14*, 1334–1337.

Fei, P., Jiao, Y.-C., Hu, W., & Zhang, F.-S. (2011). A miniaturized antipodal Vivaldi antenna with improved radiation characteristics. *IEEE Antennas and Wireless Propagation Letters, 10*, 127–130.

Ghasr, M. T., Abou-Khousa, M. A., Kharkovsky, S., Zoughi, R., & Pommerenke, D. (2012). Portable real-time microwave camera at 24 GHz. *IEEE Transactions on Antennas and Propagation, 60*, 1114–1125.

Hshieh, Y.-L., Huang, N.-T., Kan, Y.-C., & Chou, H.-T. (2013). *Improvement of far field patterns of antipodal Vivaldi antennas with step impedance resonator*. In 2013 IEEE Antennas and Propagation Society International Symposium (APSURSI). IEEE, pp. 884–885.

Karahan, M., & Armagan Sahinkaya, D. S. (2014). *A reduced size antipodal Vivaldi antenna design for wideband applications*. In 2014 IEEE Antennas and Propagation Society International Symposium (APSURSI). IEEE, pp. 1588–1589.

Kharkovsky, S., & Zoughi, R. (2007). Microwave and millimeter wave nondestructive testing and evaluation-overview and recent advances. *IEEE Instrumentation and Measurement Magazine, 10*, 26–38.

Kharkovsky, S., Case, J., Ghasr, M., Zoughi, R., Bae, S., & Belarbi, A. (2012). *Application of microwave 3D SAR imaging technique for evaluation of corrosion in steel rebars embedded in cement-based structures*. Review of Progress in Quantitative Nondestructive Evaluation, Vol. 31. AIP Publishing, pp. 1516–1523.

Kharkovsky, S., Ghasr, M., Ratnayake, R., & Percy, B. (2014). *Microwave imaging with a 3-axis multifunctional scanning system*. In Proceedings of the IEEE International Instrumentation and Measurement Technology Conference (I2MTC), pp. 1572–1575.

Kim, S.-G., & Chang, K. (2004). *Ultra wideband exponentially-tapered antipodal Vivaldi antennas*. National radio science meeting, pp. 2273–2276.

Kota, K., & Shafai, L. (2011). Gain and radiation pattern enhancement of balanced antipodal Vivaldi antenna. *Electronics Letters, 47*, 1.

Li, E., Wang, C., & Guo, G. (2016). Radiation enhanced Vivaldi antenna with double-antipodal structure. *IEEE Antennas and Wireless Propagation Letters, 16*, 561–564.

Molaei, A., Kaboli, M., Mirtaheri, S. A., & Abrishamian, M. S. (2014). Dielectric lens balanced antipodal Vivaldi antenna with low cross-polarisation for ultra-wideband applications. *IET Microwaves, Antennas and Propagation, 8*, 1137–1142.

Moosazadeh, M., & Kharkovsky, S. (2016). A compact high-gain and front-to-back ratio elliptically tapered antipodal Vivaldi antenna with trapezoid-shaped dielectric lens. *IEEE Antennas and Wireless Propagation Letters, 15*, 552–555.

Nassar, I. T., & Weller, T. M. (2015). A novel method for improving antipodal Vivaldi antenna performance. *IEEE Transactions on Antennas and Propagation, 63*, 3321–3324.

Natarajan, R., George, J. V., Kanagasabai, M., & Shrivastav, A. K. (2015). A compact antipodal Vivaldi antenna for UWB applications. *IEEE Antennas and Wireless Propagation Letters, 14*, 1557–1560.

Pastorino, M. (2010). *Microwave imaging*. Hoboken, NJ: Wiley.

Puskely, J., Lacik, J., Raida, Z., & Arthaber, H. (2016). High gain dielectric loaded Vivaldi antenna for Ka band applications. *IEEE Antennas and Wireless Propagation Letters, 15*, 2004–2007.

Quintero, G., Zurcher, J.-F., & Skrivervik, A. K. (2011). System fidelity factor: A new method for comparing UWB antennas. *IEEE Transactions on Antennas and Propagation, 59*, 2502–2512.

Teni, G., Zhang, N., Qiu, J., & Zhang, P. (2013). Research on a novel miniaturized antipodal Vivaldi antenna with improved radiation. *IEEE Antennas and Wireless Propagation Letters, 12*, 417–420.

Walden, M. C. (2013). *A wideband, 5–50+ GHz tapered-slot antenna For use in handheld test equipment*. In 2013 IEEE Antennas and Propagation Society International Symposium (APSURSI). IEEE, pp. 430–431.

Zoughi, R. (2000). *Microwave non-destructive testing and evaluation principles*. New York: Springer.

Chapter 4
Microwave and Millimetre Wave Antipodal Vivaldi Antenna with Periodic Slit-Edge Technique and the Trapezoid-Shaped Dielectric Lens for Imaging of Concrete-Based Composite Materials

4.1 Introduction

To investigate low-loss materials as well as loss materials, lower end of frequency is required. For this purpose, a combination of increasing dimensions of antenna and slits is applied in this chapter. High gain and F-to-B ratio, and wide operating bandwidth for the application in UWB continuous imaging system, are presented. It is based on the design of a CAVA that has a relatively high cut-off frequency, and its gain is low at lower frequencies. In the proposed antenna, the cut-off frequency is lowered by a factor of two by a periodic slit-edge (PSE) technique, thereby creating a periodic slit-edge AVA (PSEAVA). The gain of the antenna at low frequencies is also improved by PSE; however, the gain of both the CAVA and PSEAVA drops at higher frequencies. To increase the gain at higher frequencies, a trapezoid-shaped dielectric lens as an extension of the substrate is designed, creating periodic slit-edge AVA with a trapezoid-shaped dielectric lens (PSEAVA-TDL). Wideband impedance bandwidth, directional radiation patterns, high gain, high F-to-B ratio and narrow *E*-plane HPBW are main features of the designed PSEAVA-TDL that may satisfy requirements for microwave and millimetre wave imaging applications.

Finally, the applicability of the proposed antenna for UWB imaging of construction materials and structures for NDT is demonstrated using three different samples including a concrete block with plasterboard sheets to highlight the high-range resolution of the proposed antenna. In the first sample, the capability of the proposed antenna for the detection and evaluation of targets such as metal discs placed at different heights on foam substrate is demonstrated. In the second sample, a practical sample that is based on the cubic concrete block was prepared and imaged. A groove was made on the top surface of the block and one steel rod inserted in the groove. In the third sample, the top of the cubic concrete block was covered by plasterboard sheets and an air gap was created by removing one sheet. The third sample is considered as a cavity in a composite structure and imaging results are provided.

M. Moosazadeh, *Antipodal Vivaldi Antennas for Microwave Imaging of Construction Materials and Structures*,
https://doi.org/10.1007/978-3-030-05566-0_4

The chapter includes three main sections. In the second section of this chapter, the purpose of using periodic slit-edge technique and trapezoid-shaped dielectric lens is illustrated. The third section provides antenna performance in free space such as radiation characteristics and time domain investigations. Section 4.4 demonstrates the workability of the proposed antenna for use in microwave and millimetre wave imaging applications.

4.2 Antenna Configuration

4.2.1 Periodic Slit-Edge Technique

Figure 4.1 shows four antennas including the CAVA (cf. Fig. 4.1a), PSEAVA with one set of slits (cf. Fig. 4.1b), PSEAVA with two sets of slits (cf. Fig. 4.1c) and PSEAVA with three sets of slits (cf. Fig. 4.1d). These antennas have been designed using RO4003C substrate with relative dielectric permittivity of 3.38, a loss tangent of 0.0027 and a thickness, t_{SUB}, of 0.508 mm. The total dimensions (i.e. substrate dimensions, $L_{SUB} \times W_{SUB} \times t_{SUB}$) are $40 \times 60 \times 0.508$ mm^3. Two fins metallised on the opposite side of the substrate (referred to as top and bottom radiators) were created from the intersection of quarters of two ellipses with length of radiator, L_1. The designed antennas are fed by microstrip line with width of 1.19 mm in order to match the 50-Ω coaxial line. The intersection of a rectangular conductor with two anti-faced quarter ellipses has created a ground plane.

According to the conventional tapered slot antenna radiation theory, the flare is the main radiation part of antennas for the whole frequency band and cut-off frequency is specified by the aperture width, W_A (Schaubert et al. 1985). The antenna

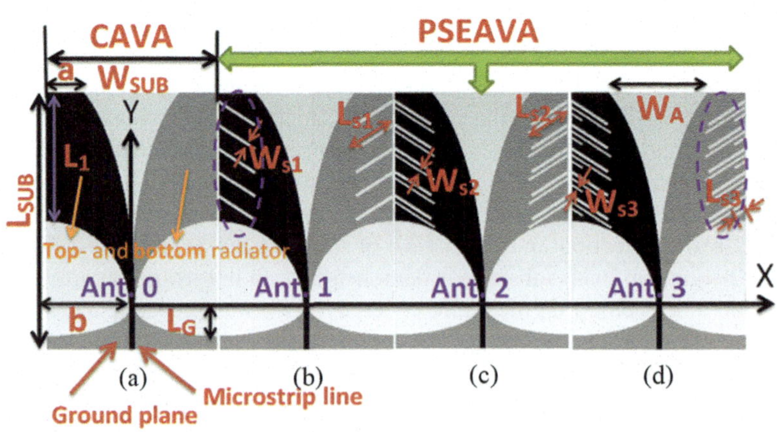

Fig. 4.1 Modification process of designed antennas: (**a**) CAVA (Ant. 0); (**b**) PSEAVA with one set of the slits (Ant. 1); (**c**) PSEAVA with two sets of the slits (Ant. 2) and (**d**) PSEAVA with three sets of the slits (Ant. 3) (Moosazadeh and Kharkovsky 2016)

cut-off wavelength can be defined as $\lambda_c = 2W_A$. In this design, the lower cut-off frequency is 6.8 GHz for the CAVA with W_A of 22.84 mm. The dimension of this CAVA has to be enlarged two times to reduce the cut-off frequency to 3.4 GHz. However, the increase of the dimension inevitably increases antenna size. To increase the operating bandwidth without significantly increasing the dimension of the antenna, the PSE technique on the radiators has been used as shown in Fig. 4.1b–d.

CST MWS is employed to optimise the antenna characteristics. For this purpose, the 3D (*XYZ*) mesh matrix of 339 × 482 × 24 is utilised with around 3.8 million mesh cells. From the parametric study on the orientation of slits, we observed that the angle of 33° with respect to *Y*-axis can provide the best performance. To explain designs of the antenna with PSE technique, the magnitude of the reflection coefficient of the antennas will be used and shown in Fig. 4.2. It can be clearly seen from Fig. 4.2a that all four antennas provide a wide operating frequency band. In order to further explain the effectiveness of the slits on S_{11}, especially at lower frequencies, frequency band between 2 and 15 GHz is shown in Fig. 4.2b. As shown in Fig. 4.2b, the lower end limitation of the CAVA (Ant. 0) for $S_{11} < -10$ dB is 6.8 GHz. PSE technique is used as follows. First, one set of the slits (Ant. 1) is applied to improve impedance matching around 5.8 GHz. Second, two sets of the slits (Ant. 2) are employed to improve impedance matching around 3.7 GHz with some degradation around 14.5 GHz. Third, three sets of the slits (Ant. 3) are used to improve impedance matching around 14.5 GHz. After an extensive simulation on the length of the periodic slits, L_{s1}, L_{s2} and L_{s3}, their optimal values were selected to be 9 mm, 9.86 mm and 2.84 mm, respectively. It should be noted that the length of the L_{s1}, L_{s2} and L_{s3} is approximately $0.25\lambda / \sqrt{\varepsilon_r}$, where λ is the corresponding wavelength of 5.2 GHz, 3.7 GHz and 14.5 GHz, respectively.

In order to further understand the behaviour of the CAVA and PSEAVA, especially at lower frequencies, surface current distribution at 3.5 and 5 GHz is shown in Fig. 4.3. It is evident that the surface current density in region A of the CAVA is very weak at 3.5 and 5 GHz, resulting in high magnitude of the reflection coefficient

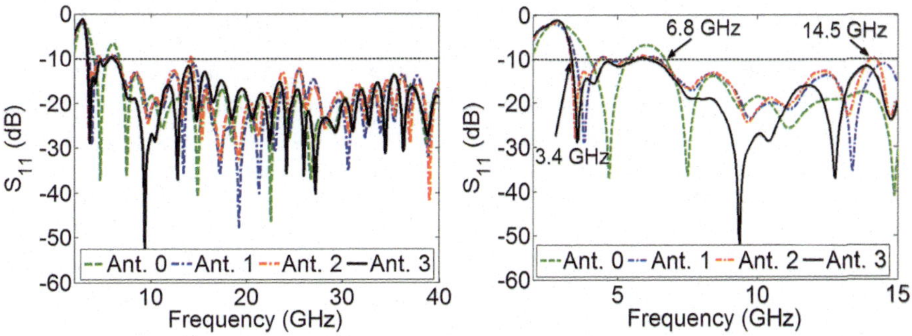

(a) (b)

Fig. 4.2 Simulated magnitude of the reflection coefficient (S_{11}) in dB of the designed antennas at (**a**) 2–40 GHz and (**b**) 2–15 GHz (Moosazadeh and Kharkovsky 2016)

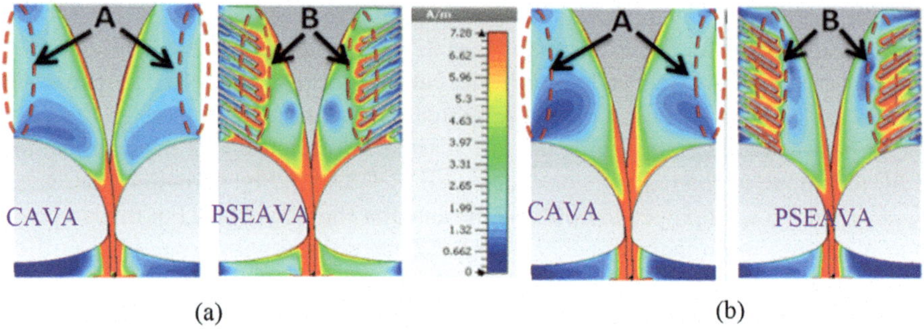

Fig. 4.3 Surface current distribution for the designed CAVA (Ant. 0) and PSEAVA (Ant. 3) at (**a**) 3.5 GHz and (**b**) 5 GHz (Moosazadeh and Kharkovsky 2016)

of the CAVA at these frequencies as shown in Fig. 4.2. On the other hand, by adding the slits, significant current density is observed in region B, indicating that the effective length of the current path is lengthened. This is the reason why the RSEAVA has the lowest cut-off frequency. The superiority of the PSEAVA in contrast with the described antenna in De Oliveira et al. (2015) is that the lower end of frequency band has been extended to 3.4 GHz without increasing dielectric constant of the substrate or length of radiators. It is worth to mention that adding slits on the outer brim of the top and bottom radiators of the antenna increases the antenna gain at lower frequencies as is shown in Fig. 4.7.

4.2.2 *Trapezoid-Shaped Dielectric Lens*

Radiation patterns of CAVA are degraded at higher frequencies because of travelling current at the termination part of radiators, resulting in a decrease of antenna gain. In order to enhance gain at higher frequencies, director or lenses have been used. The directors or lenses focus energy toward the aperture centre. In Bourqui et al. (2010) and Juan et al. (2013) employed director and lens are with higher permittivity than that of the antenna substrate. Half-elliptical shaped extension used in Wang et al. (2012) and Molaei et al. (2014) is not effective as some energy scatters away from the aperture, especially at higher frequencies. Therefore, a suitable shape of director or lens is needed to provide effective focusing. In the proposed antenna, an extension of the antenna substrate as the trapezoid-shaped dielectric lens is applied. Figure 4.4a shows a CST model of the PSEAVA-TDL where L_L and α are the length and angle of the lens, respectively. A parametric study on the length and the angle was performed and the best values were selected as 30 mm and 78.71°, respectively. It is evident that the highest value of gain is achieved at $L_L \geq 30$ mm as shown in Fig. 4.4b.

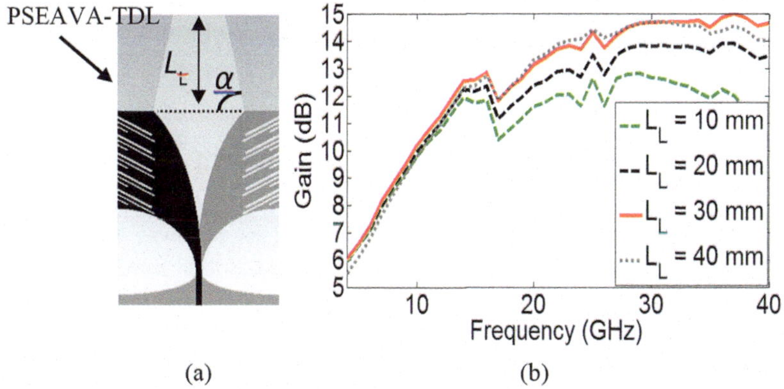

Fig. 4.4 (**a**) Schematic of the PSEAVA-TDL and (**b**) variation of the gain with frequency for the different values of L_L (Moosazadeh and Kharkovsky 2016)

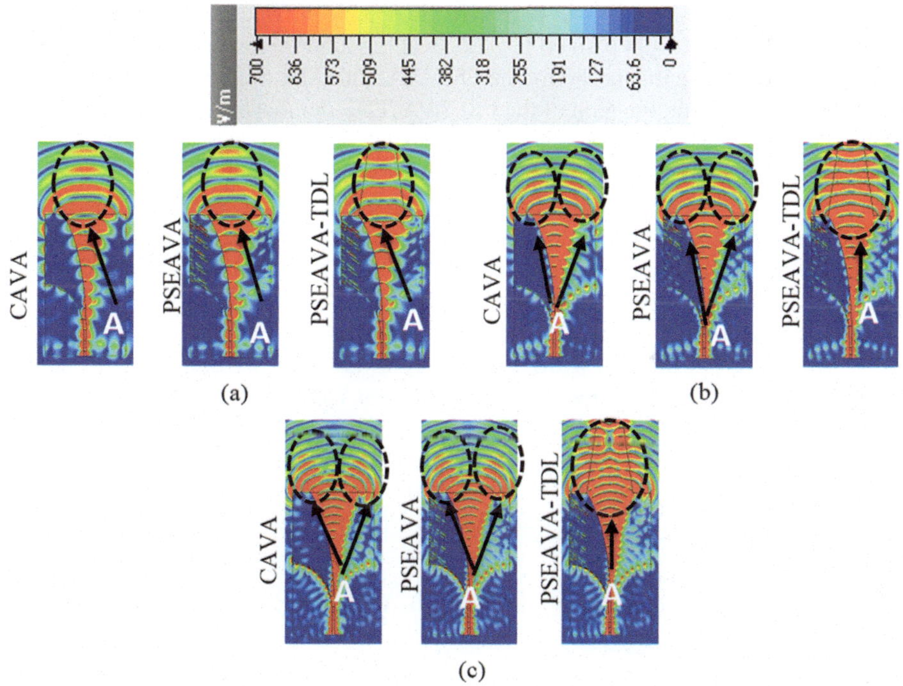

Fig. 4.5 Electric field intensity distribution with linear scale of the colour map for the designed antennas at (**a**) 20 GHz, (**b**) 30 GHz and (**c**) 40 GHz (Moosazadeh et al. 2016)

Figure 4.5 compares the electrical field intensity distribution in the near-field region of the designed antennas at three frequencies. It is clearly seen in Fig. 4.5 that the lens in PSEAVA-TDL focused the electric field in the desired direction along

Y-axis (region A) at all three frequencies while the electric field distribution for both CAVA and PSEAVA has split at 30 and 40 GHz, as shown in Fig. 4.5b, c.

4.3 Performance of the Proposed Antennas in Free Space

The optimal dimensions of the PSEAVA-TDL are as follows: W_{SUB} = 40 mm, L_{SUB} = 60 mm, a = 8.58 mm, $b = W_{SUB}/2 - W_f/2$, L_G = 6.80808 mm, W_f = 1.19 mm, W_A = 22.84 mm, L_{s1} = 9 mm, W_{s1} = 0.8 mm, L_{s2} = 9.8621 mm, W_{s2} = 0.6 mm, L_{s3} = 2.84615 mm, W_{s3} = 0.6 mm, L_1 = 30 mm and L_L = 30 mm.

The gain variation of the CAVA, PSEAVA and PSEAVA-TDL is shown in Fig. 4.6. As shown in Fig. 4.6a, gain of the CAVA has been improved at lower frequencies by slits in the PSEAVA. The gain variation of the CAVA, PSEAVA and PSEAVA-TDL at the entire frequency range is shown in Fig. 4.6b, indicating that gain of the CAVA and PSEAVA has a decreasing trend after 25 GHz. The decreasing trend of the gain of PSEAVA at higher frequencies can be attributed to the longer effective length of current path on the slits, leading to more loss at high frequencies. In order to overcome this problem, the PSEAVA with TDL (i.e. PSEAVA-TDL) has been used. As a result, the gain of the PSEAVA-TDL demonstrates rising trend after 25 GHz and its maximum value is 7.5 dB as shown in Fig. 4.6b and the PSEAVA-TDL has the highest gain (14.6 dB) at the highest frequency of its bandwidth. It is worth to mention that the designed PSEAVA with size of 40 × 60 mm and dielectric constant of the substrate of 3.38 has higher gain than referenced Vivaldi antennas which size and dielectric constant are 36.30 × 59.81 mm^2 and 6.15 (De Oliveira et al. 2015), and 50 × 56 mm^2 and 4.4 (Pandey et al. 2015). In addition, the gain at 4 GHz and 12 GHz, respectively, of the antennas is 4 dB and 8 dB for De Oliveira et al. (2015) and 3 dB and 4 dB for Pandey et al. (2015) while the gain of the PSEAVA at corresponding frequencies is 6 dB and 10.5 dB.

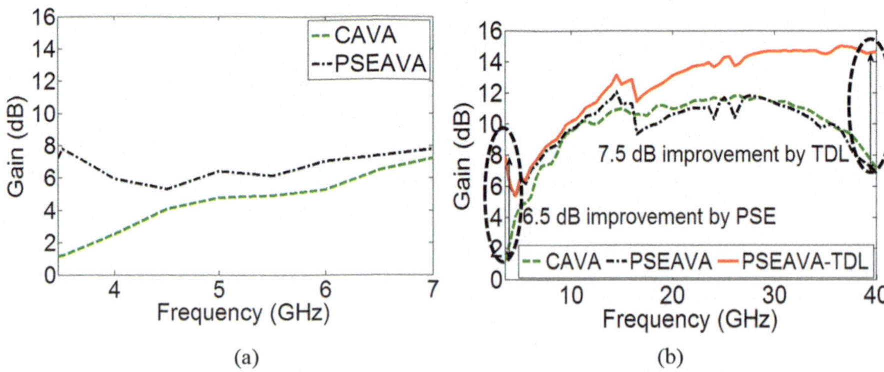

Fig. 4.6 Gain variation of (**a**) two antennas at lower frequencies (3.4–7 GHz) and (**b**) three antennas at the entire frequency range (3.4–40 GHz) (Moosazadeh and Kharkovsky 2016)

Figure 4.7 reveals the simulated radiation efficiency versus frequency from 3.4 to 40 GHz. It can be clearly seen from Fig. 4.7 that the radiation efficiency is from 87 to 97% for all three antennas.

The radiation patterns of the CAVA, PSEAVA and PSEAVA-TDL at 5, 9, 13, 19, 27 and 40 GHz for *H*-plane (*YZ*-plane) and *E*-plane (*XY*-plane) are shown in Fig. 4.8

Fig. 4.7 Simulated radiation efficiency of the designed antennas at a bandwidth of 3.4–40 GHz (Moosazadeh et al. 2016)

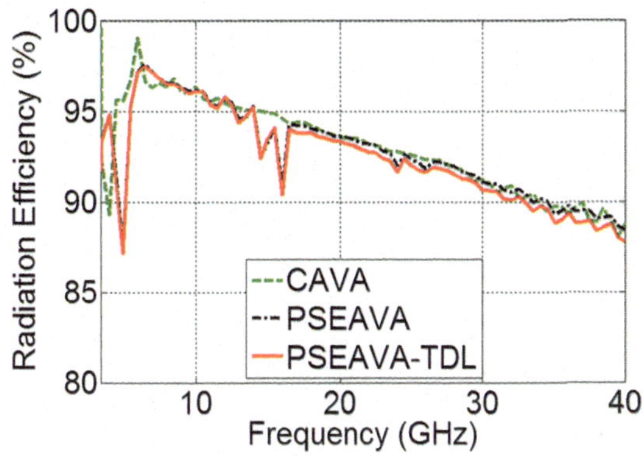

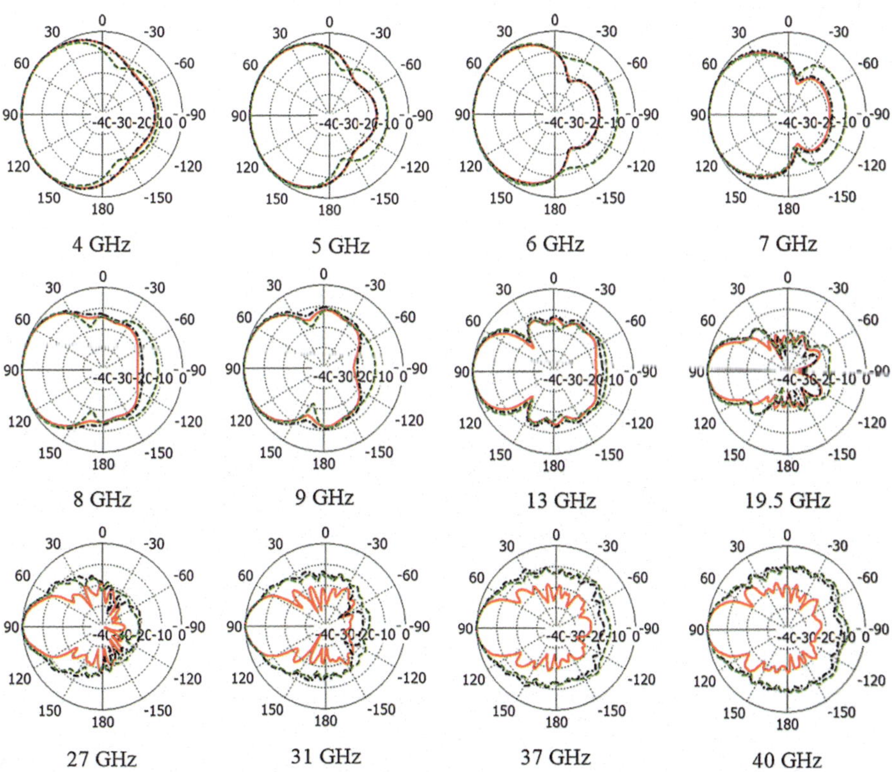

Fig. 4.8 *H*-plane radiation patterns of the antennas at different frequencies with dashed lines for CAVA, dashed dotted lines for PSEAVA and solid lines for PSEAVA-TDL (scale: −40 dB, division: 10 dB) (Moosazadeh and Kharkovsky 2016)

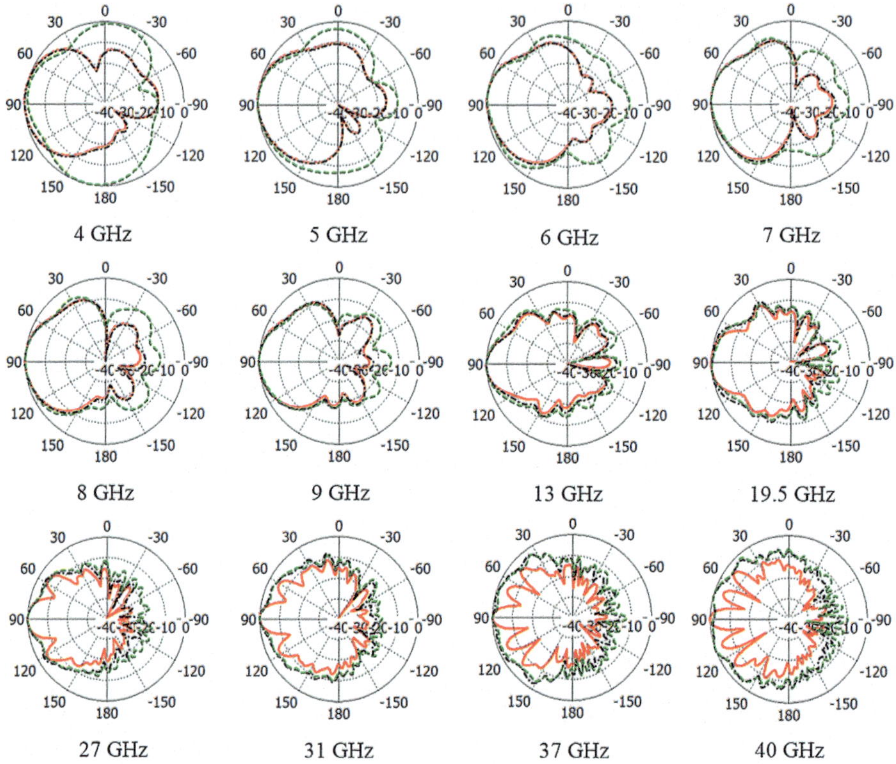

Fig. 4.9 *E*-plane radiation patterns of the antennas at different frequencies with dashed lines for CAVA, dashed dotted lines for PSEAVA and solid lines for PSEAVA-TDL (scale: −40 dB, division: 10 dB) (Moosazadeh and Kharkovsky 2016)

and Fig. 4.9, respectively. As the radiation patterns reveal, the proposed antennas have endfire characteristics with the main lobe in the axial direction (*Y*-direction in Fig. 4.1). At lower frequencies, for instance, at 5, 9 and 13 GHz, antennas have a similar performance in terms of directivity with some improvement in F-to-B ratio.

The *E*-plane 3-dB beamwidths at 40 GHz for the CAVA and PSEAVA are 64.5° and 61.5°, respectively, while it is 11.7° for the PSEAVA-TDL. The *H*-plane 3-dB beamwidths for the CAVA and PSEAVA are 40.88° and 39.83°, respectively, while it is 31.26° for the PSEAVA-TDL at 40 GHz. According to the radiation patterns, the significant enhancement in directivity within the frequency band from 19 to 40 GHz and lower side lobe level at higher frequencies can be achieved for the PSEAVA-TDL compared with the CAVA and PSEAVA.

In order to emphasise the focusing property of the PSEAVA-TDL, a summary of the obtained results for the HPBW and the tilt of beam at *E*-plane of all designed antennas is represented in Table 4.1. From given results in Table 4.1, it is necessary to point that reducing the tilt of beam is more prominent for the PSEAVA-TDL than that for the CAVA and PSEAVA at higher frequencies. Moreover, a comparison

Table 4.1 *E*-plane HPBW and tilt of beam for all designed antennas (Moosazadeh et al. 2016)

Frequency, GHz	*E*-plane HPBW, degree			*E*-plane tilt of beam, degree		
	CAVA	PSEAVA	PSEAVA-TDL	CAVA	PSEAVA	PSEAVA-TDL
6	82.7	61.7	61.7	8	5	5
7	55.4	60.2	59.7	5	4	4
9	45.3	49.9	49.9	1	0	0
13	38.4	31.2	31.2	−3	−2	−2
16	44.7	59.8	41.2	−4	−3	−3
19	39.7	55.4	42.2	−5	−11	−4
20	43.3	49.5	36.9	−7	−9	−3
21	41.5	46	33.8	−6	−8	−3
22	36.6	42.9	31.4	−4	−6	−2
23	34.9	41.3	29.7	−4	−7	−2
25	33.4	30.5	24.4	−3	−2	−1
30	24.1	25.4	18.7	0	0	0
38	67.4	67.8	12.3	−2	15	−1
39	69.8	63.2	12.2	−3	16	−1
40	64.5	61.5	11.7	−20	−21	0

among the CAVA, PSEAVA and PSEAVA-TDL at 40 GHz reveals a HPBW of 11.7° and tilt of beam of 0° in the PSEAVA-TDL, in contrast with HPBW of 64.5° and tilt of beam of −20° in the CAVA and HPBW of 61.5° and tilt of beam of −21° in the PSEAVA. In addition, the PSEAVA-TDL has narrower *E*-plane HPBW compared with some of the referenced Vivaldi antennas. For instance, the *E*-plane HPBW in Teni et al. (2013) is 20° at 30 GHz, while it is 18.7° for the PSEAVA-TDL at the same condition in terms of using lens. For the CAVA the tilt of beam is −7° at 20 GHz and −20° at 40 GHz while it is −3° at 20 GHz and 0° at 40 GHz for PSEAVA-TDL. It is valuable to point that the proposed PSEAVA-TDL is considerably capable of reducing *E*-plane beam tilts in comparison with the reported antennas in (Kota and Shafai 2011; Molaei et al. 2014). For instance, the tilt angle at 15 GHz is 4°, while it is 2° for the PSEAVA-TDL.

In addition, creating slits on the outer brim of the proposed antenna does improve the F-to-B ratio at operating bandwidth. Figure 4.10 demonstrates the F-to-B ratio changes with frequency for the CAVA, PSEAVA and PSEAVA-TDL. The maximum F-to-B ratio (37.5 dB) is achieved with the PSEAVA-TDL at a frequency of 19.5 GHz which is twice that for the CAVA at the same frequency. The magnitude of transmission coefficient (S_{21}) in dB of the designed antennas facing each other in *Y*-direction (Fig. 4.1) with 200 mm separation has been investigated. Figure 4.11 shows that the TDL increases the magnitude of transmission coefficient between 13 and 40 GHz. This fact also confirms higher directivity of the proposed antenna compared to the CAVA and PSEAVA.

The time domain behaviour of the proposed PSEAVA-TDL was investigated as well. First, in order to estimate time signal distortion, group time delay parameter of the PSEAVA-TDL was studied in *Y*-direction. To do this, two identical antennas

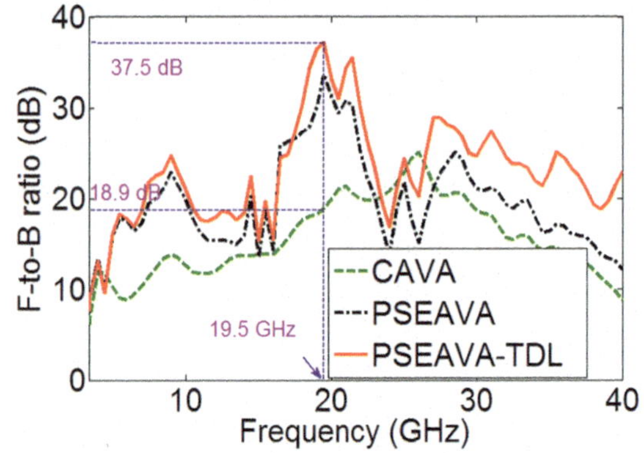

Fig. 4.10 Variation of the F-to-B ratio versus frequency of designed antennas at a bandwidth of 3.4–40 GHz (Moosazadeh and Kharkovsky 2016)

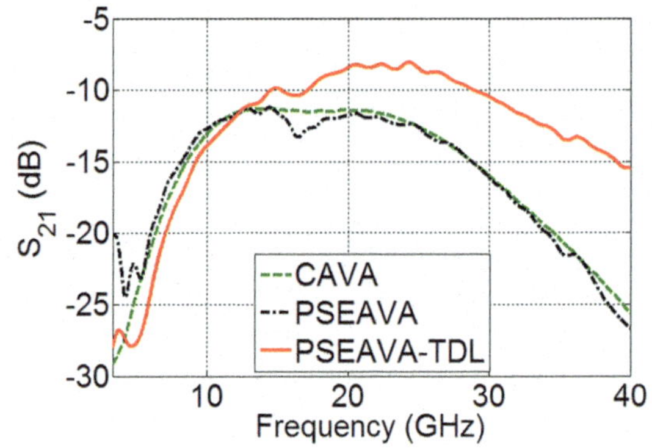

Fig. 4.11 Transmission coefficient (S_{21}) in dB of the designed antennas facing in *Y*-direction with 200 mm separation at a bandwidth of 3.4–40 GHz

were placed face to face with a distance of 200 mm to fulfil transmitting and receiving signals. It can be clearly seen from Fig. 4.12a that the maximum group time delay of the proposed antenna is around 1.5 ns with a fluctuation of less than 0.1 ns in most frequencies. In other words, this antenna is able to transmit and receive signals without serious distortion. The fidelity factor which is specified by the absolute value of the maximum of the cross-correlation coefficient, i.e. the level of similarity between transmitted and received signals, is shown in Fig. 4.12b. As shown, the maximum cross-correlation is achieved as 0.9; it means that the designed antenna operates a nearly distortion-free transmitted signal.

The performances of this antenna are also summarised in Table 4.2. Table 4.2 illustrates the specification and performances of the referenced Vivaldi antennas in terms of size, operation bandwidth, relative permittivity, and gain and F-to-B ratio in comparison with the proposed PSEAVA-TDL.

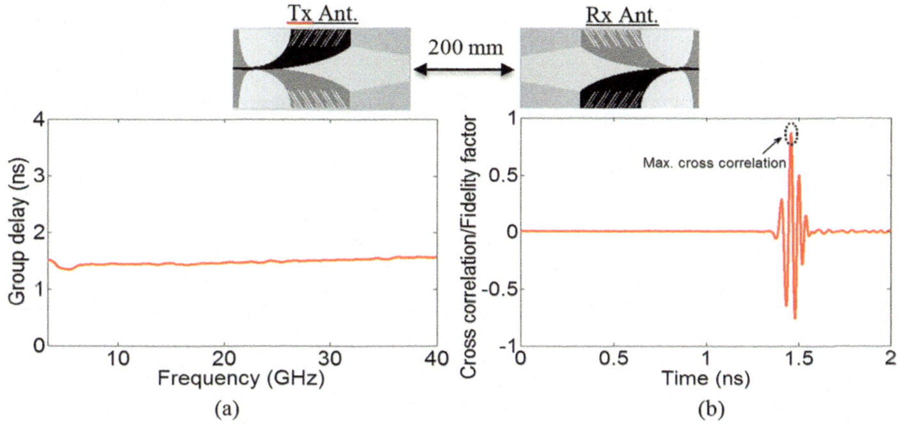

Fig. 4.12 (**a**) Group delay of the proposed PSEAVA-TDL facing in *Y*-direction with 200 mm separation at a bandwidth of 3.4–40 GHz and (**b**) cross-correlation/fidelity factor for the transmitted and received signals of the proposed PSEAVA-TDL

Table 4.2 The specification and performances of the referenced Vivaldi antennas and proposed PSEAVA-TDL (Moosazadeh and Kharkovsky 2016)

References	Dimensions (mm³)	O.B.W. (GHz)	Relative permittivity	Freq. (GHz) → Gain, F-to-B ratio (dB)
Puskely et al. (2016)	8 × 59 × 0.5	25–40	2.33	25 → 10.2, N.M. 42 → 12, N.M.
Wu et al. (2014)	41 × 48 × 0.8	3–15.1	4.4	3.5 → 5, 10 14 → 5.5, 10
In et al. (2012)	52 × 145 × 0.508	3.1–10.6	2.2	3 → 3.5, N.M 11 → 12.5, N.M.
Bai et al. (2011)	64 × 61 × 0.254	4–50	2.3	4 → 3, N.M 50 → 12, N.M.
Fei et al. (2011)	48 × 60 × 1	2.4–14	4.6	2.4 → 3.8, N.M 10 → 9, 8
Chuang et al. (2016)	17.77 × 30.48 × 0.8	9.24–18.14	3.55	N.M.
Bourqui et al. (2010)	44 × 80 × 9.2	2.4–18	2.94	N.M.
Juan et al. (2013)	70 × 140 × 3	1.5–15	2.2	3 → 8, 12.5 10 → 10, 15
Wang et al. (2012)	50.6 × 101 × 1.5	6–18	3.38	6 → 8.8, 28 16 → 10.3, 16
Kota and Shafai (2011)	56 × 105 × 3	6–20	2.5	6 → 9, N.M 20 → 11, N.M
Teni et al. (2013)	50 × 66.4 × 1	4–30	4.5	4 → 5, 13 30 → 5, 22
Molaei et al. (2014)	50 × 166 × 3.15	3–18	2.55	3.5 → 2, 7 18 → 12, 17
Proposed antenna, PSEAVA-TDL	40 × 90 × 0.508 with lens	3.4-more than 40	3.38	3.4 → 8, 9.4 19.5 → 12.8, 37.5 40 → 14, 23

'O.B.W.' and 'N.M.' state the operating bandwidth and not mentioned

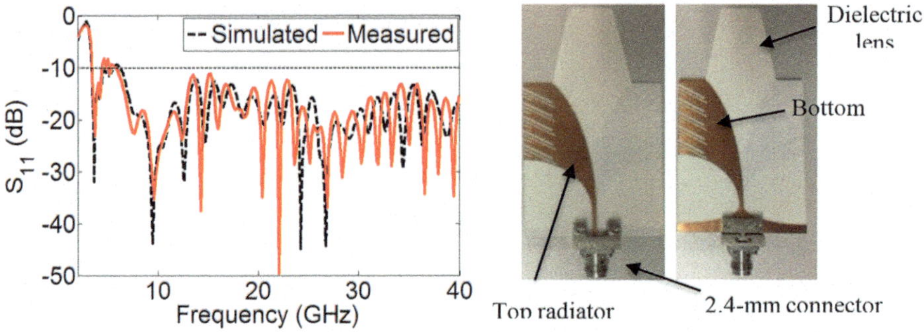

(a) (b)

Fig. 4.13 (**a**) Simulated and measured reflection coefficient versus frequency of the PSEAVA-TDL and (**b**) photograph of the fabricated antenna with 2.4 mm connector (Moosazadeh and Kharkovsky 2016)

4.4 Microwave and Millimetre Wave Imaging with the Fabricated PSEAVA-TDL

4.4.1 Fabricated Antenna and Measurement Imaging Approach

In order to validate the simulated performance of the PSEAVA-TDL at these parameters and the applicability of the proposed antenna for microwave and millimetre wave imaging applications, a prototype of the PSEAVA-TDL is fabricated and tested. The magnitude of the reflection coefficient was measured with PNA. As can be seen from Fig. 4.13a, the measured and simulated results are in good agreement at the bandwidth from 3.4 to 40 GHz. Top and back views of the fabricated PSEAVA-TDL with 2.4 mm connector are depicted in Fig. 4.13b.

To demonstrate workability and performance of the proposed PSEAVA-TDL for wideband imaging applications and to highlight range resolution of the antenna in the case of construction materials and structures, several samples have been prepared and imaged.

4.4.2 Imaging Results for the Proposed Antenna with Four Metal Discs

To demonstrate the utility and range resolution of the proposed antenna for wideband imaging applications, first a relatively simple sample panel with strong targets in free space was prepared and imaged. Figure 4.14a shows the schematic of the panel which consisted of four metal discs; each disc had a diameter of 28 mm and thickness of 2 mm. One metal disc was placed on a 200 mm by 300 mm foam

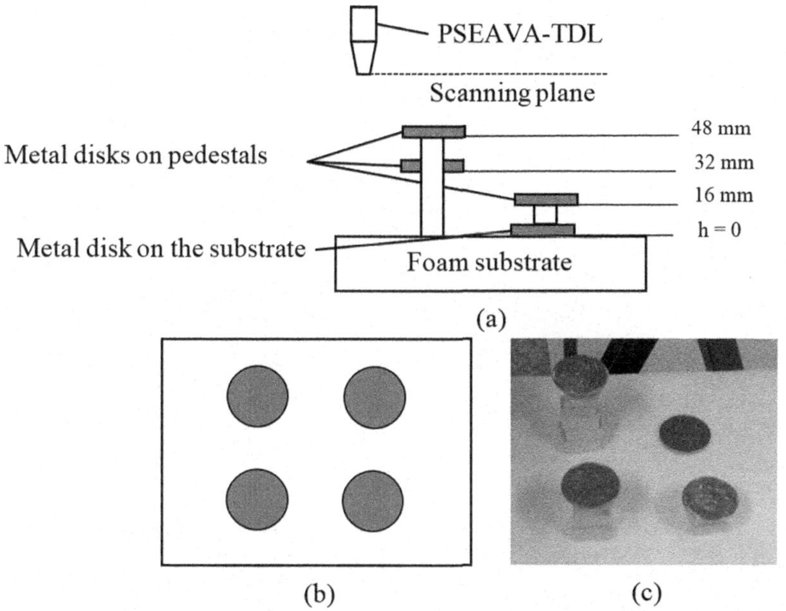

Fig. 4.14 Schematic of the panel (not-to-scale): (**a**) Schematic (side view), (**b**) schematic (top view), and (**c**) picture (perspective view) (Moosazadeh et al. 2016)

substrate at a standoff distance of ~75 mm. Other three discs were placed on foam pedestals (with a cross section of 10 mm by 10 mm) of different heights (16, 32 and 48 mm) above the foam substrate as shown in Fig. 4.14a. This difference between the locations of the discs (16 mm) was selected to be longer than the theoretical range resolution of ~4 mm for the bandwidth of 36.6 GHz. The spacing between the centres of the discs was ~60 mm as shown in Fig. 4.14b. The picture of an entire panel is shown in Fig. 4.14c. The panel was scanned with a 3 mm step size over the scanned area of 150 mm by 150 mm, while the complex reflection coefficient was measured at 101 frequency points over a frequency range of 3.4–40 GHz and 3D image of the panel was generated (Case et al. 2011).

To demonstrate the range resolution a side view of the 3D image of the panel is shown in Fig. 4.15a. It demonstrates indications of all four discs at a range of 48 mm. Fig. 4.15b–e shows horizontal slices of the 3D image (i.e. holographic slices) focused at the plane of location of each disc. The images show a clear indication of each individual disc as compared to Fig. 4.14b, c. Finally, single-frequency images over the entire frequency range were obtained and analysed. Figure 4.15f shows single-frequency 2D images of the panel at frequencies of 3.4, 9.6, 25.7 and 40 GHz which were selected to represent images of different parts of the bandwidth. An indication of all four discs as bright and dark rings is visible on the images at 9.6 and 25.7 GHz while the indication of the discs is not visible on the image at 3.4 GHz and it is not prominent on the image at 40 GHz. As a result, Fig. 4.15 demonstrates

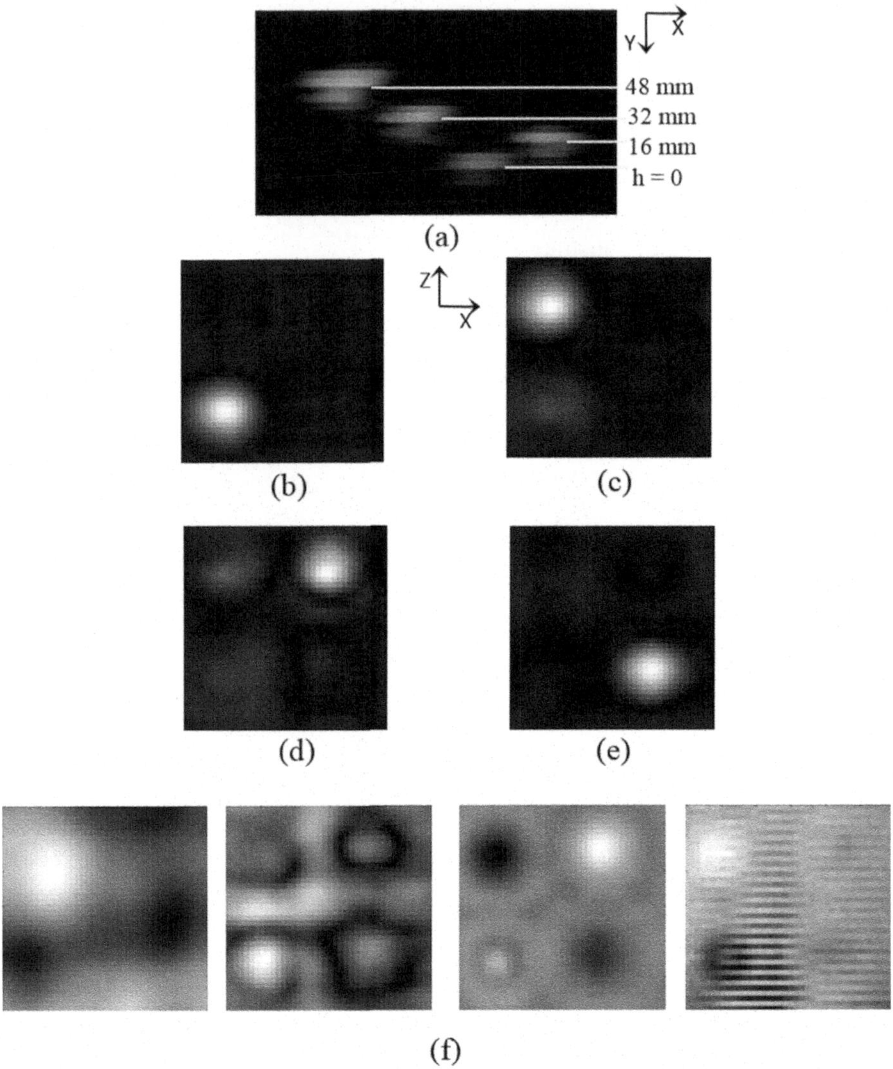

Fig. 4.15 UWB images of the panel: (**a**) 3D image (side view) and holographic slice at (**b**) $h = 48$ mm, (**c**) $h = 32$ mm, (**d**) $h = 16$ mm, (**e**) $h = 0$ mm and (**f**) 2D images without image processing at single frequencies (from left to right): 3.4, 9.6, 25.7 and 40 GHz (Moosazadeh et al. 2016)

that an acceptable indication of strong targets such as metal discs in free space on the single-frequency images and, probably, SAR images with the proposed antenna can be obtained at a frequency range from ~10 to 30 GHz. Figure 4.15 also clearly shows that the single-frequency images are more disturbed and noisy than the wide-band SAR images that are effectively a coherent sum of the single-frequency images.

4.4.3 *Imaging of Concrete Block Possessing Steel Rod in Created Groove*

In this section, to show range resolution of the proposed antenna for the wideband imaging application, a practical sample that is based on a 250 mm cubic concrete block was prepared and imaged. A groove was made on the top surface of the block; it included two parts with dimensions of 70 mm by 60 mm by 20 mm and 90 mm by 25 mm by (20–25) mm as shown in Fig. 4.16a, b (referred to as Sample 1). Another sample (referred to as Sample 2) included a 20 mm diameter steel rod inserted in the groove as shown in Fig. 4.19a.

- **Simulation results for the proposed antenna and concrete block possessing groove and steel rod in the groove**

Reflection and propagation of microwave and millimetre wave signals from and in, respectively, the concrete block with a groove and steel rod in the groove have been investigated and the results of this investigation are presented in this subsec-

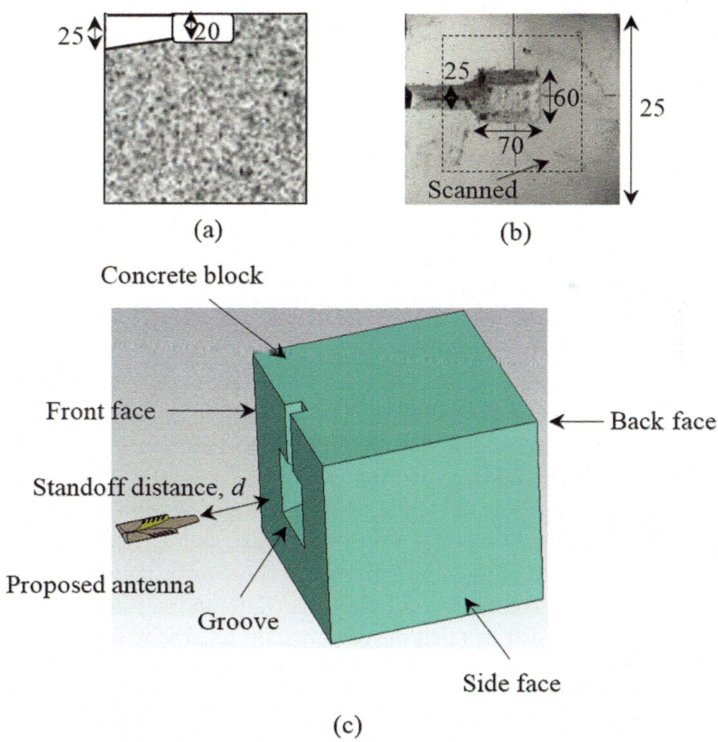

Fig. 4.16 Concrete block and Sample 1: (**a**) cross-sectional side view of the block showing the groove, (**b**) picture of the top surface of block with a groove and (**c**) model of the proposed antenna with the concrete block possessing groove in CST

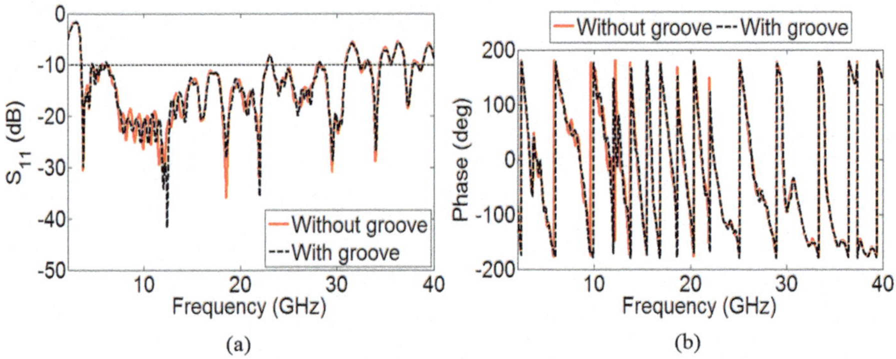

Fig. 4.17 Simulated (**a**) magnitude and (**b**) phase of the reflection coefficient with and without groove in the concrete block when the antenna was located above the centre of a wide part of the groove

tion. Firstly, simulation with a concrete block possessing a groove is performed. Secondly, simulation results for the sample with steel rod in the groove are presented. Figure 4.16c displays a model of the proposed antenna placed at the distance D from the centre of the front face of a 250 mm cubic concrete block possessing groove in CST.

Figure 4.17a, b shows simulated magnitude and phase of the reflection coefficient for dry concrete (dielectric constant of 4.5 and loss tangent of 0.0125) over the entire frequency range (3.4–40 GHz) when the antenna was located above the centre of the groove at a standoff distance of 110 mm. It can be clearly seen from Fig. 4.17 that there is a difference between values of the magnitude as well as values of the phase at lower frequencies for the concrete block with and without groove and this difference decreases when the frequency increases.

Electrical field intensity distribution for the proposed antenna and Sample 1 with and without groove at frequencies of 9.6 and 18.6 GHz is depicted with colour bar in Fig. 4.18. Zoom-in sections of Fig. 4.18 are selected to highlight the influence of groove on the field distribution. The following observation can be made from Fig. 4.18. It can be observed that the proposed antenna mainly radiates microwave signal toward Sample 1. There are standing wave patterns within the concrete block as a result of the superposition of the transmitted wave and waves reflected from the front, back and side faces of the concrete block. There are changes of electric field distribution at the interface between free space (air) and surface of the concrete block. Figure 4.18 demonstrates that the presence of the groove in the surface of the concrete block significantly changes the electric field distribution at the interface between free space and concrete block, in particular, at a frequency of 9.6 GHz. The influence of the interface between the free space and the concrete block on the electric field distribution is dominant at higher frequencies; therefore, the disturbance of this distribution by the void located at the interface is lower at higher frequencies.

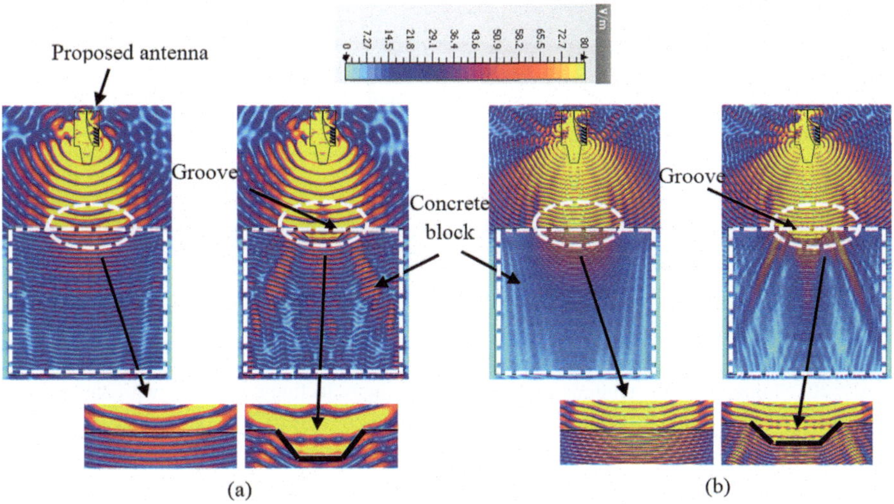

Fig. 4.18 Electrical field intensity distribution in the proposed antenna and concrete block at the frequencies of (**a**) 9.6 GHz and (**b**) 18.6 GHz: (left) without groove and (right) with groove

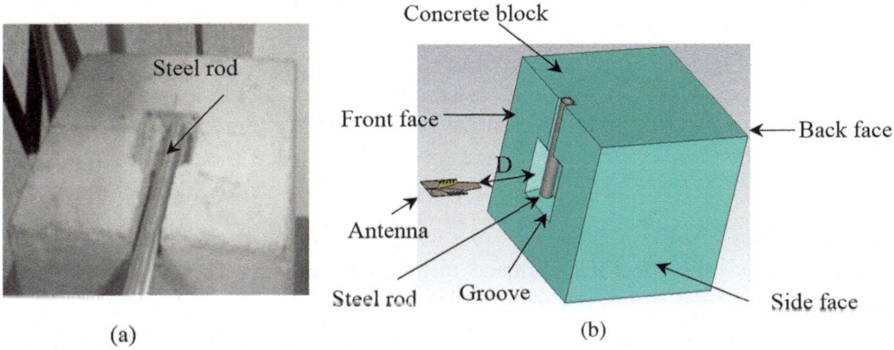

Fig. 4.19 (**a**) Picture of the concrete block with a steel rod in the groove (Sample 2) and (**b**) model of the proposed antenna with the concrete block possessing a steel rod in the groove in CST (Moosazadeh et al. 2016)

This observation explains the behaviour of the magnitude and phase of reflection coefficient at different frequencies, as shown in Fig. 4.17.

Figure 4.19a shows a picture of the concrete block with a steel rod in the groove (Sample 2). Figure 4.19b displays a model of the proposed antenna irradiating the centre of the front face of a concrete block possessing a 12 mm diameter steel rod in the groove. Figure 4.20a, b illustrates simulated magnitude and phase of the reflection coefficient for dry concrete over the entire frequency range (3.4–40 GHz)

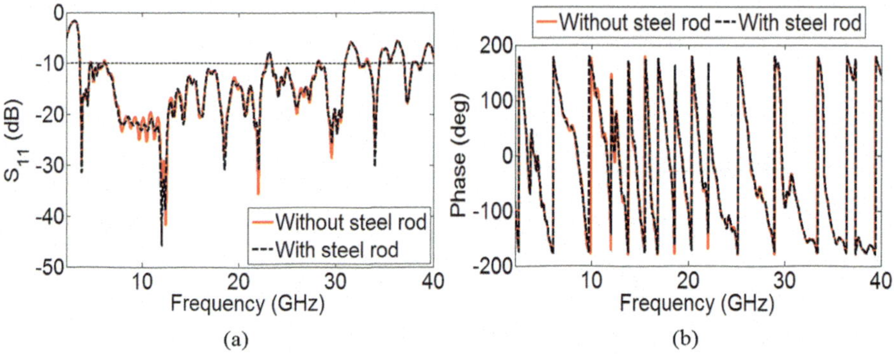

Fig. 4.20 Simulated (**a**) magnitude and (**b**) phase of the reflection coefficient with and without steel rod in groove

when the antenna was located above the centre of groove with and without steel rod at a standoff distance of 110 mm. It can be clearly seen from Fig. 4.20 that there is a difference between values of the magnitude as well as values of the phase for the concrete block with and without steel rod.

Electrical field intensity distribution for the proposed antenna and Sample 2 with and without steel rod at frequencies of 3.4, 12.03, 18.5 and 35 GHz is depicted with colour bar in Fig. 4.21. Zoom-in sections of Fig. 4.21 are selected to highlight the influence of the steel rod on the field distribution. Figure 4.21 demonstrates that the presence of steel rod in the groove of the concrete block significantly changes the electric field distribution at the interface between free space and concrete block. As shown, the presence of steel rod splits electromagnetic waves at higher frequencies, in particular, at a frequency of 18.5 GHz (c.f. Fig. 4.21c).

- **Imaging results for the proposed antenna and concrete block possessing groove and steel rod in the groove**

Samples 1 and 2 were investigated when its top surface was irradiated by the antenna at the entire frequency band at a relatively large standoff distance (distance between the antenna and top surface) of about 110 mm. Fig. 4.22a, b demonstrates the perspective and side views of the 3D image of the sample. Indications of the top of the sample and groove, as well as indications of the rod and the bottom of the groove, can be clearly seen in these images. Fig. 4.22c, d shows two holographic slices of the 3D image focused to the location of the top (c.f. Fig. 4.22c) and the bottom (c.f. Fig. 4.22d) of the groove. Indications of both the rod and the groove are clearly visible in these images. It should be noted that the rod is a relatively strong target; therefore, its indication is stronger than the indication of the groove's edges in Fig. 4.22c while the indication of the rod is weaker than the indication of the bottom of the groove in Fig. 4.22d since the image shown in Fig. 4.22d was obtained by synthetically focusing the microwave signal at the location corresponding to the bottom of the groove using the SAR technique.

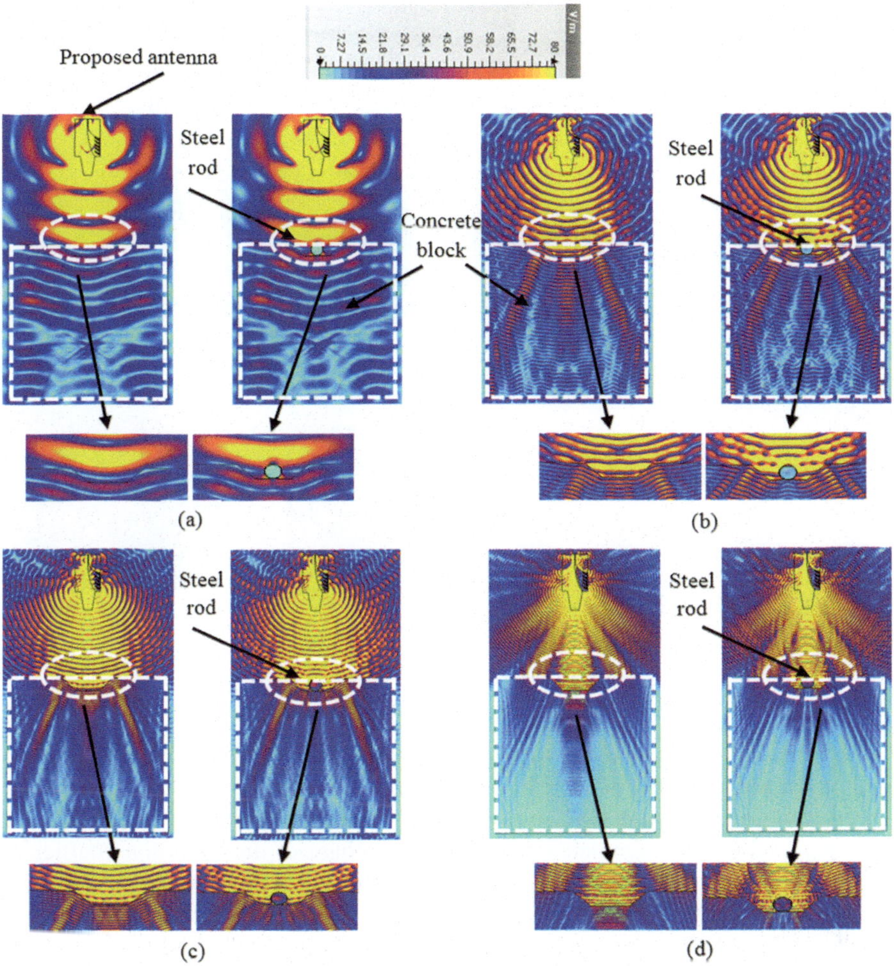

Fig. 4.21 Electrical field intensity distribution in the proposed antenna and concrete block (left) without steel rod and (right) with steel rod at the frequencies of (**a**) 3.4 GHz, (**b**) 12.03 GHz, (**c**) 18.5 GHz and (**d**) 35 GHz

4.4.4 Imaging for Concrete Block Possessing Air Gap in Plasterboard Sheets

In the third sample (referred to as Sample 3), the top of the concrete block was covered by plasterboard sheets (thickness of each sheet is ~15 mm) with ~15 mm air gap which was made by removing one sheet as shown in Fig. 4.23a, b. In this sample, the groove is considered as a cavity in a composite structure.

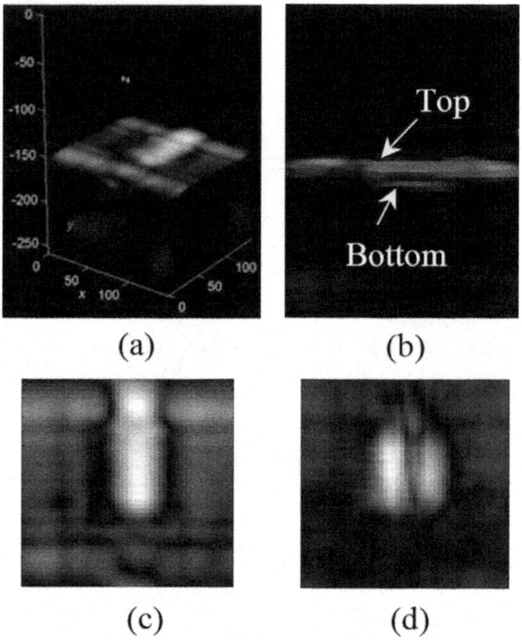

Fig. 4.22 3D image of the concrete block with a steel rod at (**a**) perspective view, and (**b**) side view and holographic slices of the concrete block with a steel rod when holographic slices focused at (**c**) the top and (**d**) the bottom of the groove (Moosazadeh et al. 2016)

To investigate reflection and transmission of electromagnetic waves in the specimen under test, a model of Sample 3 possessing air gap embedded in plasterboard sheets has been created in CST (c.f. Fig. 4.23c) and simulation was performed.

Figure 4.24 shows simulated magnitude and phase of the reflection coefficient for case of dry concrete and air gap in plasterboard sheets over the entire frequency range (3.4–40 GHz) when the antenna was located above the centre of plasterboard sheets at a standoff distance of 20 mm. It can be clearly found from Fig. 4.24 that there is a difference between values of the magnitude as well as values of phase of reflection coefficient for the concrete with and without air gap at the frequency range from 3.4 to ~13 GHz, i.e. at lower frequencies. It means that signals with frequencies within the frequency range from 3.4 to 13 GHz provide the main contribution to the detection of the air gap due to a higher penetration depth of the signals at these frequencies than at higher frequencies.

Electrical field intensity distribution for the proposed antenna and Sample 3 with and without air gap at the frequencies of 3.6, 12.6 and 30 GHz is shown in Fig. 4.25. Zoom-in sections of Fig. 4.25 are selected to highlight the influence of air gap and cavity in composite construction structures on the field distribution. Figure 4.25 demonstrates the significant disturbance of the electric field at the interface between free space (air) and plasterboard sheets. As shown, electromagnetic waves propagate in Sample 3 with multiple reflections from its boundaries and splitting of the main lobe along their propagation. Zoom-in section of Fig. 4.25 also shows changes of the electric field distribution at all three frequencies when the air gap presents

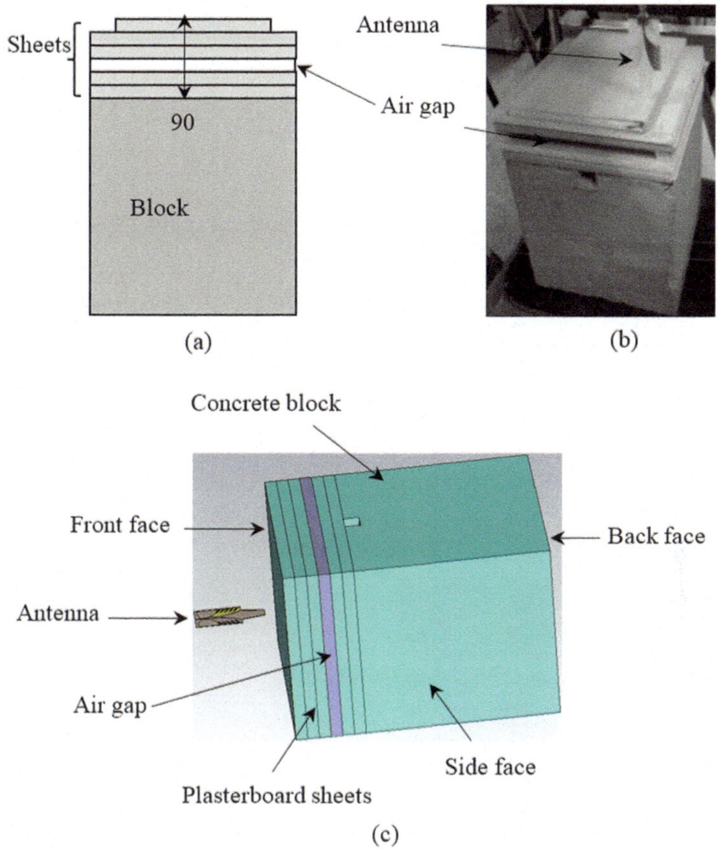

Fig. 4.23 Concrete block and sample: (**a**) schematic of the block with plasterboard sheets and air gap (**b**) picture of Sample 3 imaged by the antenna at a distance of 10 mm and (**c**) model of the proposed antenna and the concrete block with air gap in plasterboard sheets in CST (Moosazadeh et al. 2016)

inside Sample 3. In addition, the comparison of field distributions at the three frequencies clearly demonstrates different values of depth of penetration in Sample 3 at different frequencies; namely, penetration depth at the higher frequency (i.e. at 30 GHz) is less than that at the lower frequency (i.e. at 3.6 GHz). The results of the field distribution explain the behaviour of the magnitude and phase of reflection coefficient at over the entire frequency band shown in Fig. 4.24.

- **Imaging results for the proposed antenna with concrete block possessing air gap in plasterboard sheets**

Figure 4.26a, b shows the perspective and side views of the 3D image of Sample 3. They clearly demonstrate the range resolution provided by the proposed antenna

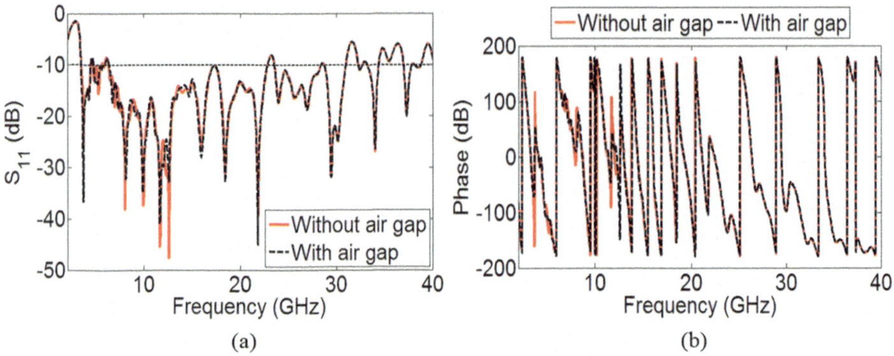

Fig. 4.24 Simulated (**a**) magnitude and (**b**) phase of the reflection coefficient with and without air gap in plasterboard sheets

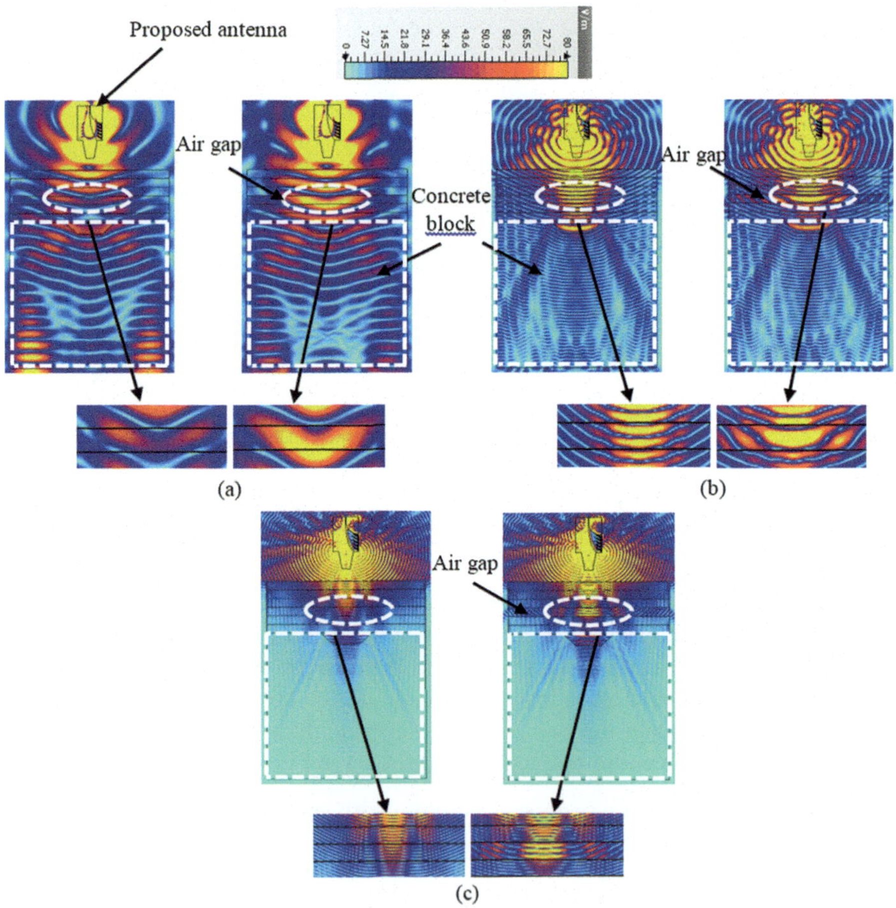

Fig. 4.25 Electrical field intensity distribution in the proposed antenna and concrete block with and without air gap in plasterboard sheets at the frequencies of (**a**) 3.6 GHz, (**b**) 12.6 GHz and (**c**) 30 GHz

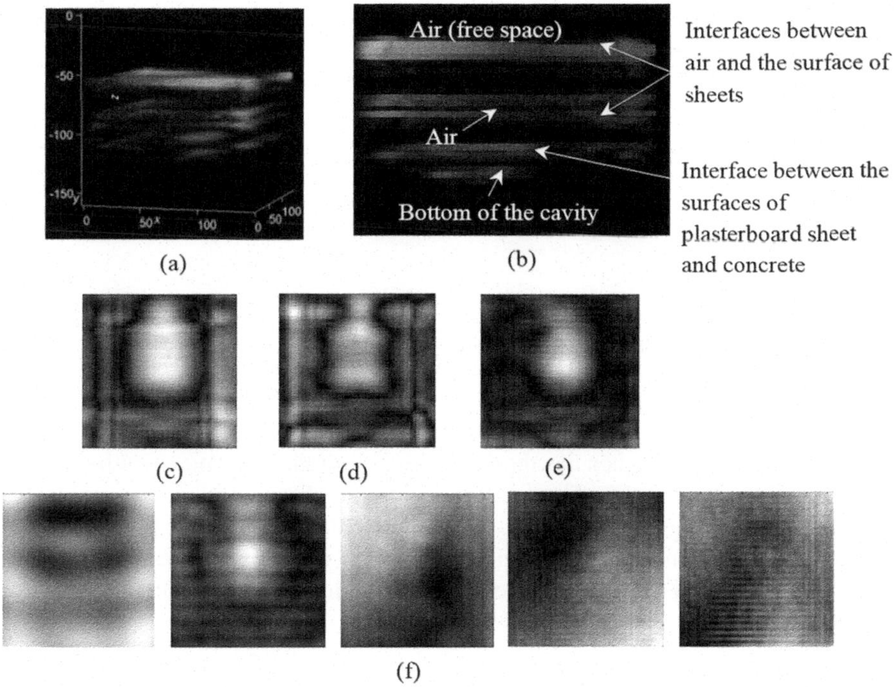

Fig. 4.26 Images and holographic slices of the concrete block with plasterboard sheets, air gap and cavity: 3D image at (**a**) perspective view and (**b**) side view; holographic slices focused at the (**c**) top, (**d**) middle and (**e**) bottom of the cavity and (**f**) 2D images of Sample 3 without image processing at single frequencies (from left to right): 3.8, 10.35, 22.07, 35.61 and 38.17 GHz (Moosazadeh et al. 2016)

in conjunction with the SAR algorithm. Thus, flaws such as air gap and cavity in composite construction structures may be detected. An indication of two interfaces between air and sheets as bright lines and air gap between the interfaces as a dark line can be clearly seen in the images of Figs. 4.26a, b. The interface between the top surface of block and the lowest surface of plasterboard sheets as well as the bottom of cavity is visible in Fig. 4.26a, b.

Images of holographic slices of the cavity are shown in Fig. 4.26c–e, for which several observations can be made corresponding to the top, middle and bottom of the cavity. First, they show the indications of different features of the cavity such as its non-uniform shape, thereby demonstrating the high-range resolution of this UWB technique. The images of the top of cavity showing the indication of its edges (c.f. Fig. 4.26c) and the bottom of cavity (c.f. Fig. 4.26e) are very similar to the images of the groove in concrete block without sheets (c.f., Fig. 4.22c, d, respectively). This observation demonstrates a good penetration capability of this technique in a plasterboard layered structure with the air gap. Additional information can be obtained from single-frequency images. For this purpose, Fig. 4.26f shows single-frequency 2D images of Sample 3 at the frequencies of 3.8, 10.35, 22.07, 35.61 and 38.17 GHz. They show that indications related to the cavity are evident through sample depth at lower frequencies (c.f. images at 3.8 and 10.35 GHz in Fig. 4.26f) while the indication of the cavity is not visible on the image at higher frequencies (c.f. images at 22.07, 35.61 and 38.17 GHz in Fig. 4.26).

4.5 Summary

High-quality imaging of construction materials and structures requires ultra-wideband and directive microwave and millimetre wave antennas with relatively high gain and front-to-back ratio. In addition, the bandwidth of the antenna should include the range of frequencies that provides a desired penetration depth. For this purpose, a novel periodic slit-edge antipodal Vivaldi antenna with a trapezoid-shaped dielectric lens was designed, tested and applied for the imaging of samples made of low-loss construction materials. The results of the investigation into performance of the designed antennas and comparison with some of the referenced Vivaldi antennas show that the proposed antenna has the highest gain (up to 16 dB) and front-to-back ratio (up to 37.5 dB), and the narrowest half-power beamwidth (down to 11.7°). A fabricated prototype of the proposed antenna with compact size of $0.45\lambda \times 1.02\lambda \times 0.005\lambda$ mm^3 (where λ is the free-space wavelength corresponding to the lowest operating frequency of 3.4 GHz) demonstrates a good agreement between measured and simulated results. Capability of the proposed antenna for imaging of samples made of construction materials with void inclusions is also demonstrated. It is shown that this antenna in conjunction with a synthetic aperture radar algorithm provides high-range-resolution images of the samples with clear indication of the inclusions. This technique can be used for non-destructive testing and evaluation of infrastructure materials and structures such as bridges and buildings.

References

Bai, J., Shi, S., & Prather, D. W. (2011). Modified compact antipodal Vivaldi antenna for 4–50-GHz UWB application. *IEEE Transactions on Microwave Theory and Techniques, 59*, 1051–1057.

Bourqui, J., Okoniewski, M., & Fear, E. C. (2010). Balanced antipodal Vivaldi antenna with dielectric director for near-field microwave imaging. *IEEE Transactions on Antennas and Propagation, 58*, 2318–2326.

Case, J. T., Ghasr, M. T., & Zoughi, R. (2011). Optimum two-dimensional uniform spatial sampling for microwave SAR-based NDE imaging systems. *IEEE Transactions on Instrumentation and Measurement, 60*, 3806–3815.

Chuang, J.-K., Cheng, Y.-H., & Wang, C.-L. (2016). *Compact and broadband microstrip-fed antenna using antisymmetric tapered probe with triangular element*. In 2016 IEEE International Symposium on RadioFrequency Integration Technology (RFIT). IEEE, pp. 1–3.

De Oliveira, A. M., Perotoni, M. B., Kofuji, S. T., & Justo, J. F. (2015). A palm tree antipodal Vivaldi antenna with exponential slot edge for improved radiation pattern. *IEEE Antennas and Wireless Propagation Letters, 14*, 1334–1337.

Fei, P., Jiao, Y.-C., Hu, W., & Zhang, F.-S. (2011). A miniaturized antipodal Vivaldi antenna with improved radiation characteristics. *IEEE Antennas and Wireless Propagation Letters, 10*, 127–130.

In, D. M., Lee, M. J., Kim, D., Oh, C. Y., & Kim, Y. S. (2012). Antipodal linearly tapered slot antenna using unequal half-circular defected sides for gain improvements. *Microwave and Optical Technology Letters, 54*, 1963–1965.

Juan, L., Guang, F., Lin, Y., & Demin, F. (2013). A modified balanced antipodal Vivaldi antenna with improved radiation characteristics. *Microwave and Optical Technology Letters, 55*, 1321–1325.

Kota, K., & Shafai, L. (2011). Gain and radiation pattern enhancement of balanced antipodal Vivaldi antenna. *Electronics Letters, 47*, 1.

Molaei, A., Kaboli, M., Mirtaheri, S. A., & Abrishamian, M. S. (2014). Dielectric lens balanced antipodal Vivaldi antenna with low cross-polarisation for ultra-wideband applications. *IET Microwaves, Antennas and Propagation, 8*, 1137–1142.

Moosazadeh, M., & Kharkovsky, S. (2016). A compact high-gain and front-to-back ratio elliptically tapered antipodal Vivaldi antenna with trapezoid-shaped dielectric lens. *IEEE Antennas and Wireless Propagation Letters, 15*, 552–555.

Moosazadeh, M., Kharkovsky, S., & Case, J. T. (2016). Microwave and millimetre wave antipodal Vivaldi antenna with trapezoid-shaped dielectric lens for imaging of construction materials. *IET Microwaves, Antennas and Propagation, 10*, 301–309.

Pandey, G., Verma, H., & Meshram, M. (2015). Compact antipodal Vivaldi antenna for UWB applications. *Electronics Letters, 51*, 308–310.

Puskely, J., Lacik, J., Raida, Z., & Arthaber, H. (2016). High gain dielectric loaded Vivaldi antenna for Ka band applications. *IEEE Antennas and Wireless Propagation Letters, 15*, 2004–2007.

Schaubert, D. H., Kollberg, E. L., Korzeniowski, T. L., Thungren, T., Johansson, J., & Yngvesson, K. S. (1985). Endfire tapered slot antennas on dielectric substrates. *IEEE Transactions on Antennas and Propagation, 33*, 1392–1400.

Teni, G., Zhang, N., Qiu, J., & Zhang, P. (2013). Research on a novel miniaturized antipodal Vivaldi antenna with improved radiation. *IEEE Antennas and Wireless Propagation Letters, 12*, 417–420.

Wang, P., Zhang, H., Wen, G., & Sun, Y. (2012). Design of modified 6-18 GHz balanced antipodal Vivaldi antenna. *Progress in Electromagnetics Research C, 25*, 271–285.

Wu, J., Zhao, Z., Nie, Z., & Liu, Q.-H. (2014). A printed UWB Vivaldi antenna using stepped connection structure between slot-line and tapered patches. *IEEE Antennas and Wireless Propagation Letters, 13*, 698–701.

Chapter 5
Antipodal Vivaldi Antenna with Regular Triangular Shaped Slits for Microwave Imaging of Concrete Materials and Structures

5.1 Introduction

An AVA with trapezoid-shaped dielectric lens (Moosazadeh and Kharkovsky 2016) has been presented in Chap. 4. However, its impedance bandwidth is restricted at lower operating frequencies (i.e. 3.4 GHz). As a result, this antenna (Moosazadeh and Kharkovsky 2016) was applied for imaging low-loss construction materials such as plasterboard sheets (Moosazadeh et al. 2016a).

An UWB AVA designed to operate at low frequency of 2 GHz and with high gain at low frequencies for microwave imaging of concrete materials and structures is presented in this chapter. First, a CAVA is designed as a reference. The operating frequency of the designed CAVA has been restricted at the low end of frequency band. Second, to extend the low end of frequency band, the inner edges of the top and bottom radiators of CAVA have been bent (referred to as AVA-B). Third, to enhance gain at lower frequencies, regular triangular slit (RTS) in saw-shaped configuration is applied to the AVA-B (referred to as RTSAVA-B). The time group delay of the RTSAVA-B is studied as well.

The chapter includes three main sections. In the second section of this chapter, a bending technique and regular triangular shaped slit technique are detailed. The third section provides antenna performance in free space. Finally, applicability of the proposed antenna for UWB imaging of construction materials and structures for non-destructive testing and evaluation of concrete specimens possessing an air-filled cavity (i.e. void) and plastic pipe is demonstrated in Sect. 5.4.

M. Moosazadeh, *Antipodal Vivaldi Antennas for Microwave Imaging of Construction Materials and Structures*,
https://doi.org/10.1007/978-3-030-05566-0_5

5.2 Proposed Antenna, Results and Discussions

5.2.1 Bending Technique

The modification process has been applied in this section using two antennas, namely CAVA and AVA-B as shown in Fig. 5.1a, b. It is well known that deeper penetration depth, $\delta_{s,}$ of electromagnetic waves inside high-loss construction materials such as concrete can be provided at lower operating frequency. In this chapter, 2 GHz is the target for the lowest operating frequency (Shaari et al. 2004). To achieve the desired penetration depth and high-range resolution, a design of the antenna is elaborated as follows. The antennas were printed on a Rogers RO4003C substrate with thickness of 0.508 mm, relative dielectric constant of 3.38 and loss tangent of 0.0027. We have used thin and low dielectric constant substrate to avoid radiation characteristic degradation at higher frequencies since increasing electrical size of thickness causes unwanted modes. According to previous research (Schaubert et al. 1985), the effective substrate thickness should meet the requirement of $0.005\lambda_0 < h_{\text{eff}} < 0.03\lambda_0$, where λ_0 is the wavelength at the centre frequency of the operating bandwidth and h_{eff} is $h\left(\sqrt{\varepsilon_r}-1\right)$. The antennas are fed through a microstrip line with a width of 1.14 mm in order to match with the 50-Ω coaxial line. The top and bottom radiators are formed by the intersection of a quarter of inner and outer edges of two ellipses with major radii (U_1 and U_2) and the secondary radii (V_1 and V_2), as shown in Fig. 5.1b. The tapered ground plane section is created from the intersection of a rectangular shaped conductor with dimensions of $W_g \times L_g$ and two anti-faced quarter circles with radius of r.

To investigate the antenna performance, full-wave electromagnetic models of the antennas are developed using CST Microwave Studio software employed to

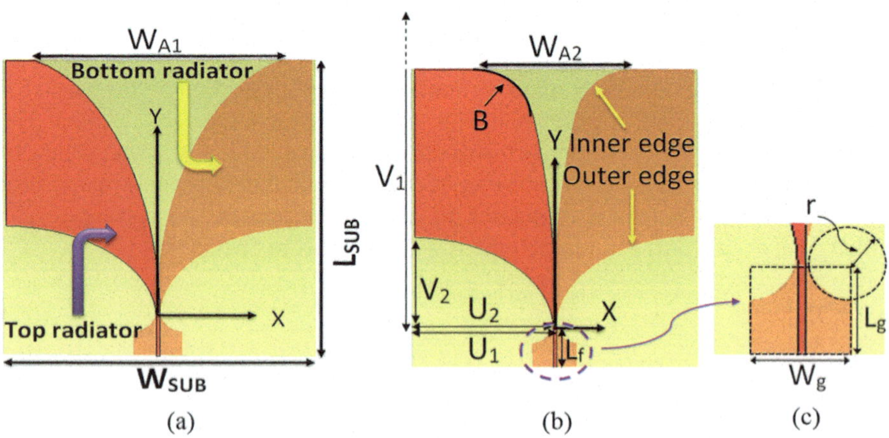

Fig. 5.1 Configuration of the designed AVAs in *XY*-plane: (**a**) CAVA, (**b**) AVA-B with opening elliptical rate, *B*, and (**c**) zoom-in of the ground plane configuration (Moosazadeh and Kharkovsky 2015a)

optimise the antenna characteristics. For this purpose the 3D (XYZ) mesh matrix of 525 × 536 × 22 is utilised with around 5.8 million mesh cells.

The simulated magnitude of reflection coefficient of the antennas for the entire operating bandwidth is plotted in Fig. 5.2a. It can be seen from Fig. 5.2a that the designed antennas have wide bandwidth, thus satisfying the requirement for high-range resolution for microwave imaging application. In order to further analyse the magnitude of reflection coefficient at lower frequencies, the frequency range from 2 to 5 GHz is depicted in Fig. 5.2b. The cut-off frequency is specified by aperture width (i.e. separation between top and bottom radiators), W_A, which is half of wavelength at the lowest operating frequency (Schaubert et al. 1985). Therefore, by taking into account $f_L \approx 2$ GHz, the AVA should have W_A of 75 mm to meet the small size of the antenna. Two steps for the designed antennas are considered.

The first option is to design CAVA with the lowest operating frequency of 2 GHz to meet UWB performance and desired penetration depth of electromagnetic waves (c.f. Fig. 5.1a). It is obviously observed that the CAVA has a bandwidth from 3.3 to 27 GHz, as shown in Fig. 5.2a, and a poor impedance matching occurs between 2.5 and 3.3 GHz, as shown in Fig. 5.2b. As a result, the W_{A1} of this antenna has to be increased to have the lowest operating frequency of 2 GHz. However, this violates our goal of small antenna size to satisfy the requirement for small size of the technique (i.e. its portability). The second option is to extend the low end of the CAVA frequency band by elliptically bending the inner edges of the top and bottom radiators (c.f. Fig. 5.1b). It should be mentioned that the elliptical bending used in this design on elliptical line of inner edges of the radiators is set from CST MWS bend option. As shown in Fig. 5.2a, the designed AVA-B has an operating bandwidth from 2.2 to 27 GHz and improves impedance matching between 2.2 and 5 GHz as shown in Fig. 5.2b without enlarging the width of the antenna.

In order to further understand the behaviour of the CAVA and AVA-B at low frequencies, E-field distribution at 3 GHz is shown in Fig. 5.3. It is clear that the intensity of E-field in radiation region 'A' of the CAVA is very weak, leading to high magnitude of the reflection coefficient of the CAVA as shown in Fig. 5.2b. By

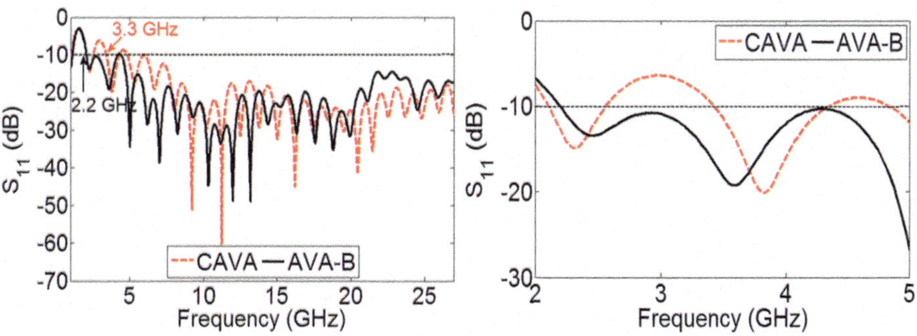

(a) (b)

Fig. 5.2 Simulated magnitude of the reflection coefficient (S_{11}) in dB for the designed antennas showed in Fig. 5.1 at (**a**) 1–27 GHz and (**b**) 2–5 GHz (Moosazadeh and Kharkovsky 2015a)

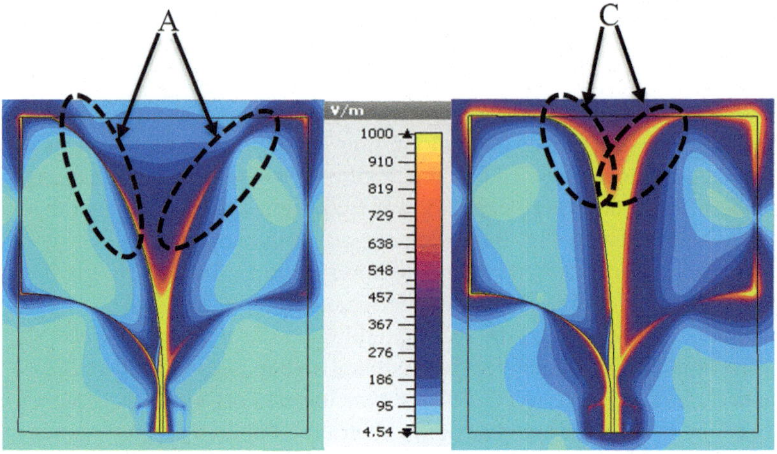

Fig. 5.3 E-field distribution in *XY*-plane of the designed CAVA and AVA-B at 3 GHz

elliptically bending inner edges of the top and bottom radiators, significant E-field concentration is achieved in radiation region 'C' of the AVA-B, resulting in lower magnitude of the reflection coefficient, as shown in Fig. 5.2b.

Substituting the estimated values for the real and imaginary parts of concrete at 2 GHz (Schaubert et al. 1985) ($\varepsilon_r' = 5.9$ and $\varepsilon_r'' = 0.6$ for moisture content of 6.2%) into Eqs. (2.5) and (2.6) gives the penetration depth of 0.2 m that is sufficient for testing many concrete structures in practice.

In the AVA-B, opening elliptical bending rate, *B*, and length of radiator, V_1 (c.f., Fig. 5.1b), are the most critical design parameters. Consequently, both parameters must be optimised to simultaneously achieve the widest impedance bandwidth and the best impedance matching. First, by maintaining *B*, we investigate the dependence of S_{11} on V_1. Figure 5.4a, b displays the magnitude of reflection coefficient of AVA-B with $B = 20$ and four values of V_1. We regard both the entire bandwidth and lower frequencies to show that the impedance matching is improved at lower frequencies when V_1 is 132.5 mm.

Second, by keeping V_1 constant at 132.5 mm, we investigated changes of *B* on impedance bandwidth. Figure 5.5a, b shows the magnitude of reflection coefficient of AVA-B with $V_1 = 132.5$ mm and four values of *B* at the entire frequency range and lower frequencies. It can be clearly seen from Fig. 5.5b that impedance matching at lower frequencies is improved with *B* of 20. Thus, the AVA-B with *B* of 20 and V_1 of 132.5 mm suggest the UWB performance with bandwidth from 2.2 to 27 GHz.

Gain variation of the proposed AVA-B is investigated as well. The gain variation at lower frequencies (2–6 GHz) for the selected *B* values is plotted in Fig. 5.6. As shown, the highest gain is achieved at *B* of 20. However, the AVA-B gain is low at lower frequencies.

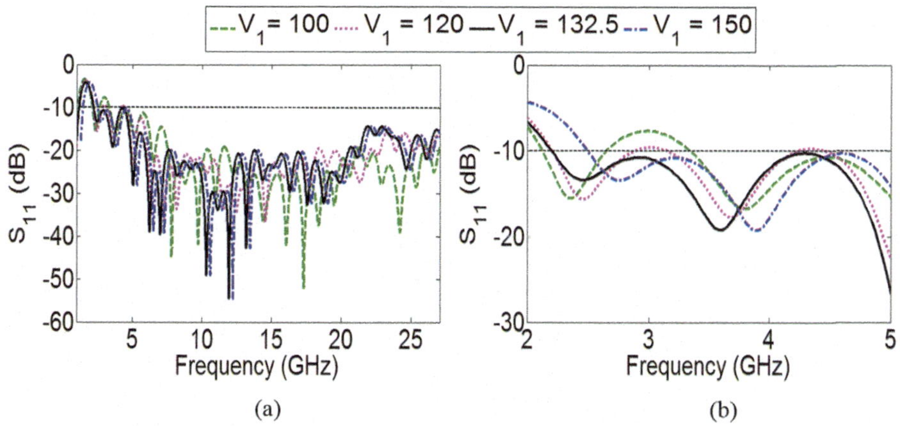

Fig. 5.4 Simulated magnitude of the reflection coefficient (S_{11}) in dB of AVA-B with $B = 20$ for four values of V_1 at bandwidths of (**a**) 1–27 GHz and (**b**) 2–5 GHz

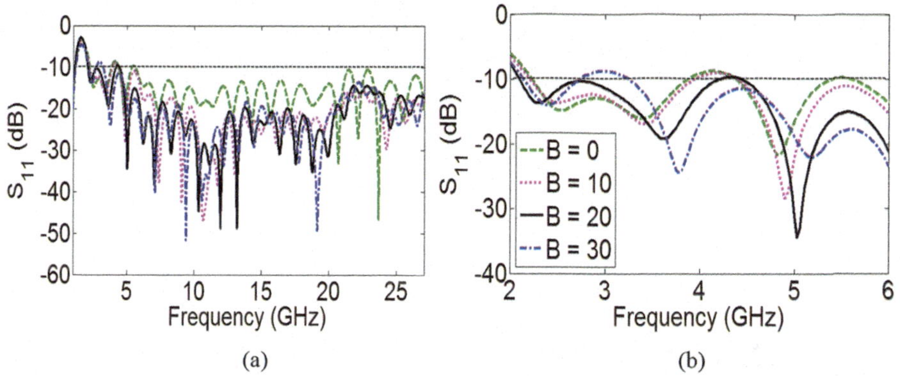

Fig. 5.5 Simulated magnitude of the reflection coefficient (S_{11}) in dB of the designed AVA-B with $V_1 = 132.5$ mm for four values of B at bandwidths of (**a**) 1–27 GHz and (**b**) 2–6 GHz

5.2.2 Regular Triangular Shaped Slit

To increase gain of the AVA-B at low frequencies, RTS in saw-shaped configuration has been applied in edges of the radiators. Figure 5.7a shows the geometry of the proposed RTSAVA-B. As a result of the parametric study, the optimal parameters of the designed RTSAVA-B are as follows: $W_{SUB} = 96$ mm, $L_{SUB} = 100$ mm, $W_{A2} = 53.19$ mm, $U_1 = (W_{SUB} + W_f)/2$, $V_1 = 132.5$ mm, $U_2 = (W_{SUB} - W_f)/2$, $V_2 = 31$ mm, $B = 20$, $X_1 = 5$ mm, $Y_1 = 2$ mm, $Y_2 = 5.5$ mm, $r = (W_g - W_f)/2$, $L_f = 13$ mm, $L_g = 15$ mm and $W_g = 15$ mm.

The variation of gain versus frequency for AVA-B and RTSAVA-B at the entire frequency range (2–27 GHz) and lower (2–5 GHz) frequencies is depicted in Fig. 5.7b. As shown in Fig. 5.7b, the RTS not only enhances gain at lower frequencies but also increases gain at higher frequencies. Also, the highest gain (peak gain of 12.45 dB at 20 GHz) is attributed to the RTSAVA-B.

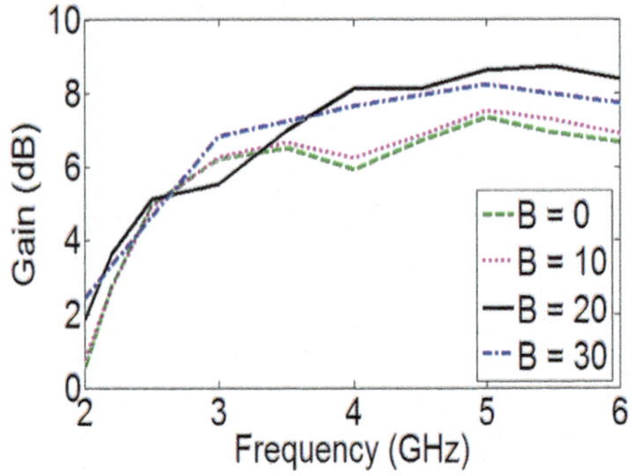

Fig. 5.6 Gain variation with frequency of the AVA-B with V_1 = 132.5 mm for four values of B at the lower frequencies (2–6 GHz)

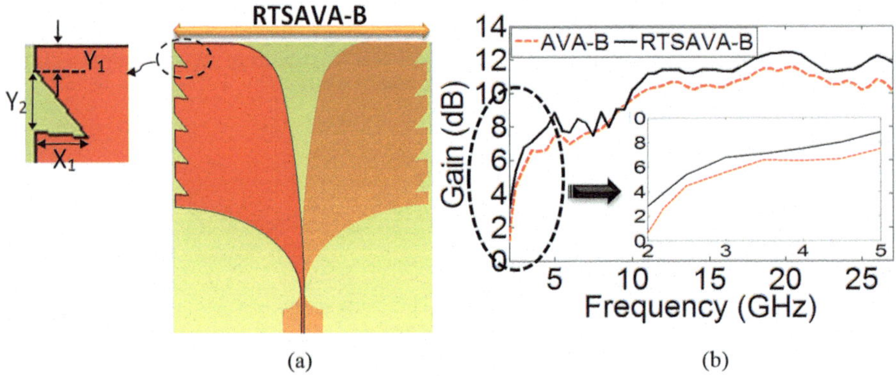

Fig. 5.7 (**a**) Configuration of the designed RTSAVA-B and (**b**) gain variation for the AVA-B and RTSAVA-B at the entire frequency range (2–27 GHz) and zoom-in for lower frequencies (2–5 GHz) (Moosazadeh and Kharkovsky 2015a)

The simulated magnitude of the reflection coefficient (S_{11}) in dB for the designed AVA-B and RTSAVA-B is plotted in Fig. 5.8a. In order to further analyse the magnitude of reflection coefficient at lower frequencies, the frequency range from 2 to 3 GHz is depicted in Fig. 5.8b. As shown in Fig. 5.8b, low end of frequency band extends up to 2 GHz for the proposed RTSAVA-B.

To characterise the designed RTSAVA-B, its gain at the bandwidth of 2.5–5 GHz is better than gain values of the referenced Vivaldi antennas as displayed in Table 5.1. As can be seen, the gain of RTSAVA-B is higher than that of the referenced Vivaldi antennas.

The radiation efficiency of the proposed RTSAVA-B is simulated and shown in Fig. 5.9. It can be seen from Fig. 5.9 that it is above 87% over whole operating frequency range and decreases when frequencies increase. The simulated H- (YZ) and E- (XY) plane radiation patterns of the designed AVA-B and RTSAVA-B at frequencies of 2.2, 7, 16, 23 and 27 GHz are depicted in Fig. 5.10. The maximum radiation occurs on Y-direction (Fig. 5.1) which demonstrates the directionality of the designed antennas. A low cross-polarisation <−15 dB is obtained in both E- and H-planes.

It should be mentioned that RTS also improves the F-to-B ratio. Figure 5.11 depicts F-to-B ratio versus frequency for the AVA-B and RTSAVA-B. As shown in Fig. 5.11, an improvement of the F-to-B ratio from 8 to 27 GHz for the designed RTSAVA-B is achieved compared to the AVA-B. The high F-to-B ratio of 37.64 dB is obtained with the RTSAVA-B at 16 GHz whereas it is 17.23 dB with the AVA-B.

To estimate time signal distortion, the group time delay behaviour of the proposed RTSAVA-B is investigated. To do this, two profiles of the RTSAVA-B are put

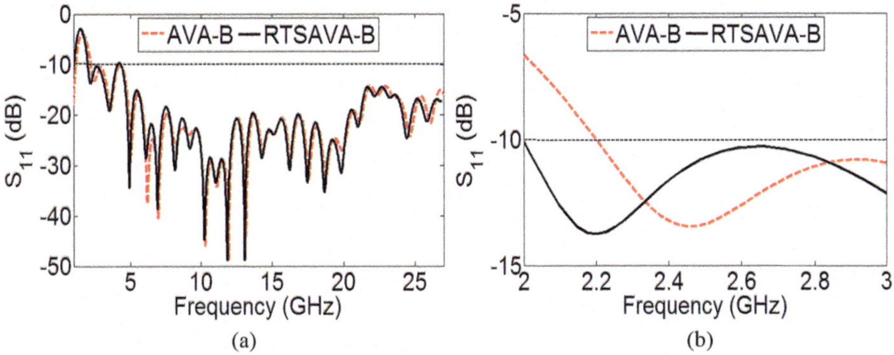

Fig. 5.8 Simulated magnitude of the reflection coefficient (S_{11}) in dB for the designed AVA-B and RTSAVA-B at bandwidths of (**a**) 1–27 GHz and (**b**) 2–3 GHz

Table 5.1 Gain of the referenced Vivaldi antennas and the proposed antenna at lower frequencies

References	Gain (dB) at lower frequencies			
	2.5 GHz	3 GHz	4 GHz	5 GHz
Zhu et al. (2012)	–	2	3.5	4
Lee et al. (2012)	–	2.5	3.5	4.5
Wu et al. (2014)	–	5.5	5	5
Pandey et al. (2015)	2.5	2.8	3	4.2
Bai et al. (2011)	1.2	1.5	2.5	4
Fei et al. (2011)	3.8	5.3	4.8	5
Teni et al. (2013)	–	–	5	6.5
Natarajan et al. (2015)	–	–	1.8	4
De Oliveira et al. (2015)	–	–	4.2	5.8
Wang et al. (2016)	3.8	4	5	6.5
Bourqui et al. (2010)	Not mentioned			
Molaei et al. (2014)	–	2.8	3.5	4
Moosazadeh and Kharkovsky (2015b)	–	–	6	6.5
Nassar and Weller (2015)	−2	0	6	8
Wang et al. (2007)	−4	2	3	3.5
Moosazadeh and Kharkovsky (2016)	–	–	6	6.5
Abbosh (2009)	–	2.7	4.2	6.5
Proposed antenna, RTSAVA-B	5.4	6.7	7.5	8.9

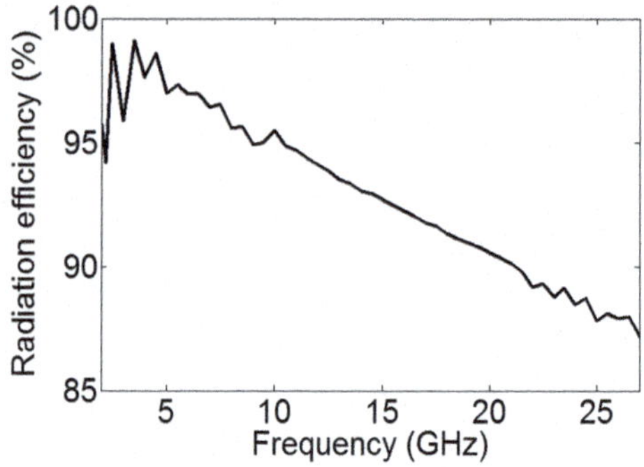

Fig. 5.9 Variation of the radiation efficiency with frequency for the designed RTSAVA-B

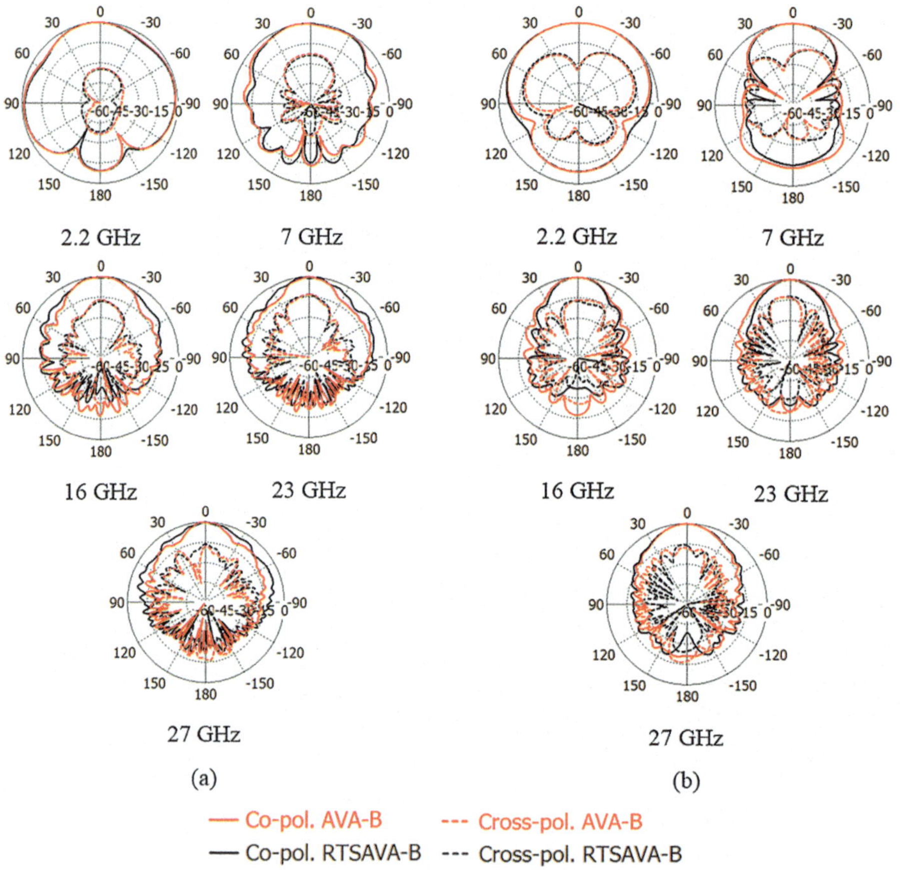

Fig. 5.10 The simulated co- and cross-polarisation of (**a**) *E*- and (**b**) *H*-plane radiation patterns of the designed AVA-B and RTSAVA-B at different frequencies

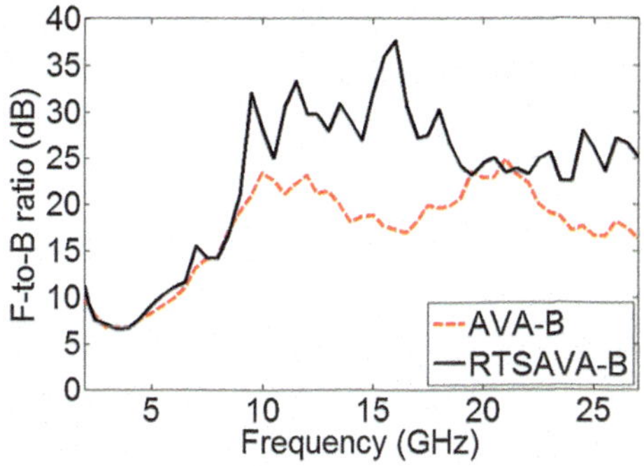

Fig. 5.11 Variation of the F-to-B ratio with frequency for the designed AVA-B and RTSAVA-B

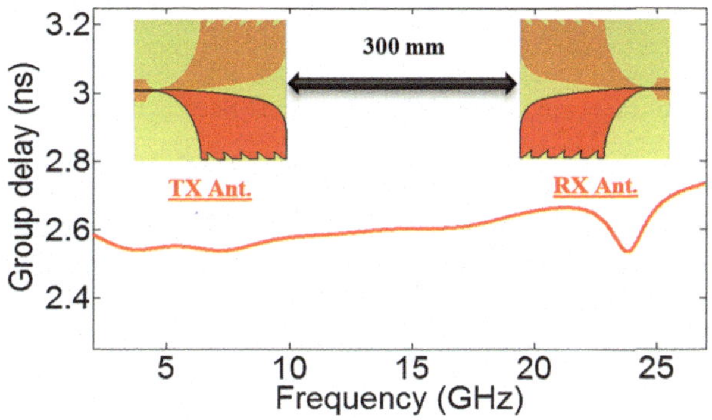

Fig. 5.12 Group delay of the proposed RTSAVA-B and two-antenna arrangement in endfire direction (inset)

in endfire direction (*Y*-direction in Fig. 5.1) with a distance of 300 mm to act as transmitting (*TX*) and receiving (*RX*) antennas as shown in Fig. 5.12 (inset). It can be clearly seen that the supreme group time delay of the proposed antenna is around 2.6 ns with fluctuation of less than 0.1 ns in most frequencies. It can be found that the proposed antenna is able to transmit and receive signals without serious distortion.

Overall, the achieved results show that the RTSAVA-B is the best candidate for microwave imaging applications.

5.3 Microwave Imaging with the Fabricated RTSAVA-B

5.3.1 Fabricated Antenna and Measurement Imaging Approach

A prototype of the RTSAVA-B with the optimal parameters was fabricated and tested with PNA to validate the simulated S_{11} results. A photo of the fabricated antenna with SMA End Launch connector is shown in Fig. 5.13a. The measured and simulated magnitude of the reflection coefficient (S_{11}) in dB of the proposed antenna is plotted in Fig. 5.13b, indicating that the antenna has a bandwidth from 2 to 27 GHz

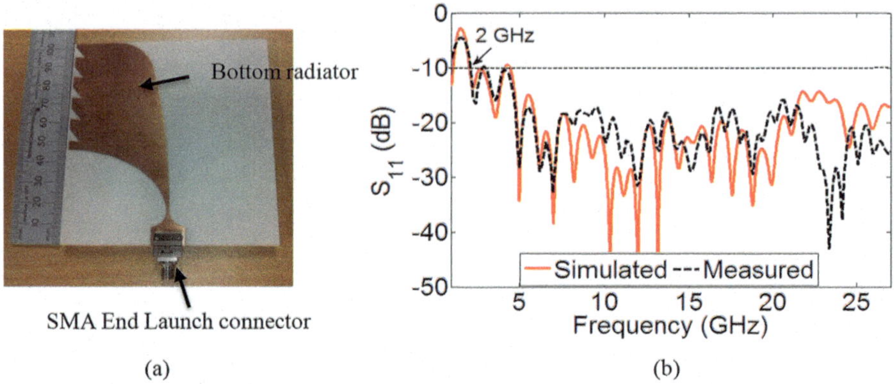

Fig. 5.13 (**a**) Photograph of the fabricated RTSAVA-B (top view) with SMA End Launch connector and (**b**) simulated and measured magnitude of the reflection coefficient (S_{11}) versus frequency of the proposed antenna (Moosazadeh and Kharkovsky 2015a)

for S_{11} <−10 dB. As shown in Fig. 5.13b, there is good agreement between measured and simulated results except for some discrepancies at middle and higher frequencies which can be due to inaccuracy in fabrication of antenna and connector.

To show the applicability of the proposed RTSAVA-B for wideband microwave imaging applications and to highlight penetration depth and range resolution of the imaging system with the antenna in the case of construction materials and structures, several specimens possessing voids as flaws or structural features such as an air-filled plastic pipe have been designed.

To study the influence of the flaws and structural features inside the specimens on the interaction between microwave signal and specimens at different frequencies within the entire frequency range, a computational model of the proposed antenna and the specimens has been created and investigated using CST MWS. Then, specimens were fabricated and imaged.

5.3.2 Proposed Antenna and the Concrete Block with Construction Foam Slab Possessing Metal Disc

In this section the results of microwave imaging and simulation for three concrete specimens are presented. Two of these specimens include a 250 mm concrete block with artificially created grooves on its top surface as shown in Fig. 5.17a. One groove was wide (referred to as groove 1) and another was narrow (referred to as groove 2) with dimensions of approximately 70 mm by 60 mm by 20 mm and 90 mm by 25 mm by 20 mm, respectively. It should be noted that the edges of grooves were not uniform and their walls were tilted. The block was covered by a 60 mm thick construction foam slab (referred to as specimen 1) as shown in Fig. 5.14a or a 100 mm thick concrete slab (referred to as specimen 2) as shown in Fig. 5.18a. When the grooves were covered by the foam or concrete slab, they were

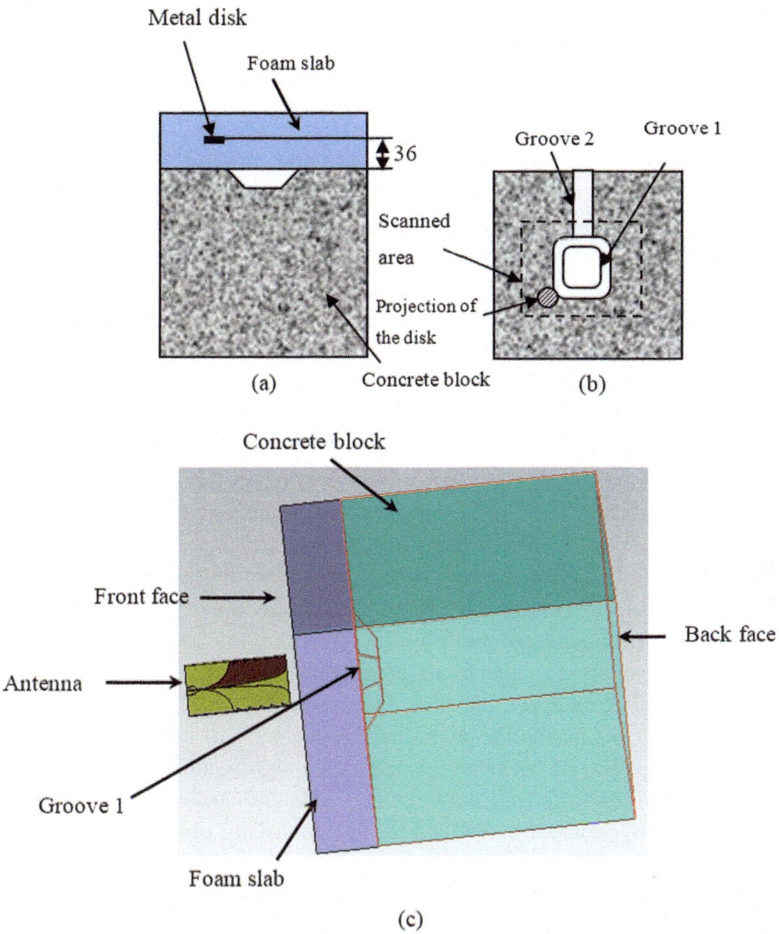

Fig. 5.14 Schematic of the concrete block with two grooves on the top surface; and cross sectional view of specimen 1: (**a**) side view, (**b**) top view of the top face of the block under the foam slab and (**c**) a model of specimen 1 in CST (Moosazadeh and Kharkovsky 2015a)

considered as cavities in a composite structure. In addition, in specimen 1 the foam slab possessed a 28 mm diameter metal disc (thickness of 2 mm) embedded at the distance of 36 mm from the surface of concrete block as shown in Fig. 5.14a, b.

Specimens 1 and 2 were scanned with a 3 mm step along each axis over the scanned area of 150 mm by 150 mm (c.f. Fig. 5.14b) at a standoff distance of 10 mm. The complex reflection coefficient was measured at 101 frequency points over a frequency range of 2–27 GHz and 3D images of the specimens were generated using the measured data and the SAR algorithm.

- **Simulation results for specimen 1**

Figure 5.14c displays a model of the proposed antenna placed at a distance of 10 mm from the centre of the front face of a 60 mm thick construction foam slab along with 250 mm cubic concrete block possessing groove 1 in CST.

Figure 5.15a, b shows simulated magnitude and phase of reflection coefficient for dry concrete (dielectric constant of 4.5 and loss tangent of 0.0125) and standard construction foam over the entire frequency range (2–27 GHz) when the antenna was located above the centre of groove 1 at a standoff distance of 10 mm. It can be seen from Fig. 5.15a that there are differences between values of the magnitude as well as values of the phase for the concrete with and without cavity/void and these differences decrease when the frequency increases.

To understand this effect and other features, electrical field intensity distribution for the proposed antenna and specimen 1 with and without void at frequencies of 3.81 and 6.73 GHz is shown with colour bar in Fig. 5.16. Zoom-in sections of Fig. 5.16 are selected to highlight the influence of cavity/void on the field distribution. There are standing wave patterns within the foam slab and the concrete block as a result of the superposition of transmitted wave and waves reflected from front, back and side faces of the concrete block. It can be observed that the proposed antenna mainly radiates microwave signal toward specimen 1 and practically there is no disturbance of the electric field by the foam slab due to a low relative (to free space) dielectric constant of the foam. However, there are remarkable changes of electric field distribution at the interface between the foam slab and concrete block even without void, in particular, at higher frequency, i.e. 6.73 GHz. Figure 5.16 shows that the presence of void inside specimen 1 significantly changes the electric field distribution at the interface between the foam slab and concrete block, in particular, at lower frequency, i.e. 3.81 GHz. The following observation can be made from Fig. 5.16. The influence of the interface between the foam slab and concrete block on the electric field distribution is dominant at higher frequencies; therefore, the disturbance of this distribution by the void located at the interface is lower at

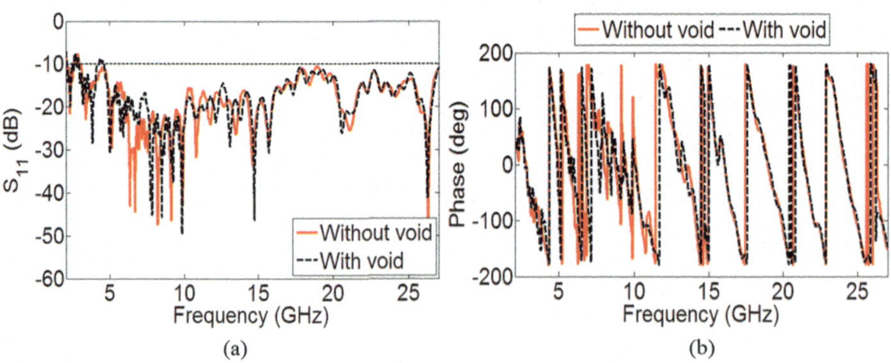

Fig. 5.15 Simulated (**a**) magnitude and (**b**) phase of the reflection coefficient versus frequency with and without void inside the concrete at a standoff distance of 10 mm

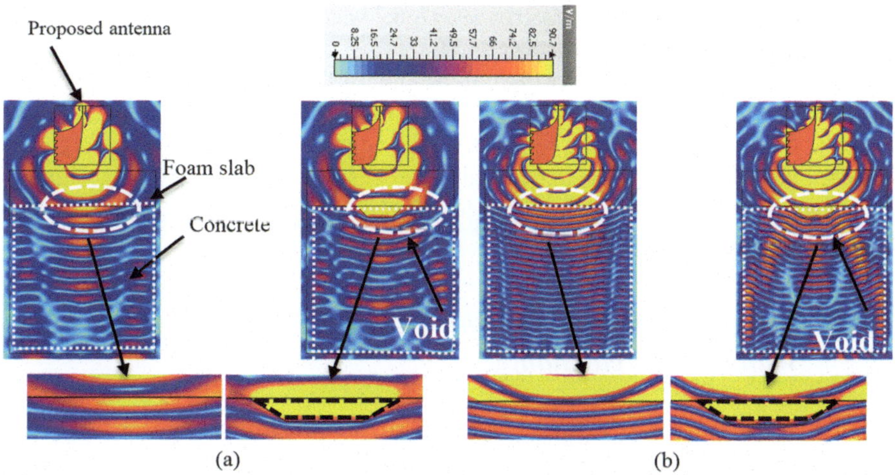

Fig. 5.16 Electrical field intensity distribution in the proposed antenna and specimen 1 (left) without and (right) with void at frequencies of (**a**) 3.81 GHz and (**b**) 6.73 GHz

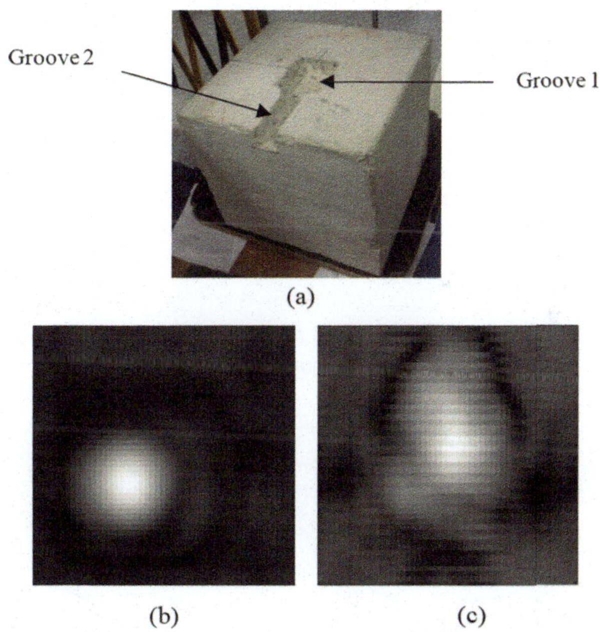

Fig. 5.17 (**a**) Picture of the concrete block with two grooves on the top surface and holographic slices of the 3D image of specimen 1 when microwave signal was focused at the location of (**b**) the disc and (**c**) the surface of concrete block (Moosazadeh and Kharkovsky 2015a)

higher frequencies. This observation explains the behaviour of the magnitude and phase of reflection coefficient at different frequencies as shown in Fig. 5.15a, b.

- **Imaging results for specimen 1**

Figure 5.17a shows a picture of the concrete block with artificially created two grooves on the top surface. Figure 5.17b, c shows holographic horizontal slice of the

3D image of specimen 1 when microwave signal was focused at the location of disc (Fig. 5.17b) and the surface of concrete block (Fig. 5.17c). The following observations can be made from Fig. 5.17b, c. Indication of the disc without any indication of the cavity is clearly visible in Fig. 5.17b. Figure 5.17c indicates the cavity that is slightly distorted by indication of the disc since the metal disc is a strong target. In general, these images demonstrate a high-range resolution of the imaging technique with the proposed antenna and SAR algorithm.

5.3.3 Proposed Antenna and the Concrete Block with a Concrete Slab

Simulation with the same arrangement of scanned area and standoff distance has been performed with specimen 2 (c.f., Fig. 5.18a). It should be noted that compared to specimen 1 this specimen does not have a contrast of dielectric permittivity between the block and the slab, and the concrete slab has higher dielectric constant, loss factor and thickness than the foam slab.

- **Simulation results for specimen 2**

Figure 5.18b displays a model of the proposed antenna placed at a distance of 10 mm from the centre of the front face of a 100 mm thick concrete slab along with 250 mm cubic concrete block possessing groove 1 in CST.

Figure 5.19a, b shows the magnitude and phase of reflection coefficient at a standoff distance of 10 mm for specimen 2, i.e. the dry concrete block and concrete slab. It shows that there is a visible difference between values of the magnitude and values of phase of reflection coefficient for the concrete with and without void at a

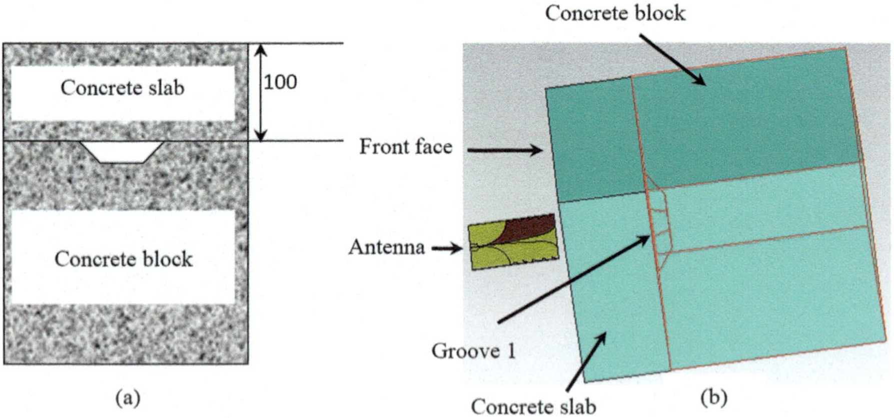

Fig. 5.18 (**a**) Cross-sectional side view of specimen 2 and (**b**) model of the proposed antenna placed at a distance of 10 mm from the centre of the front face of a 100 mm concrete slab along with 250 mm cubic concrete block possessing groove 1 in CST

frequency range from 2 to ~14 GHz, i.e. at lower frequencies. It means that signals with frequencies within a frequency range from 2 to 14 GHz mainly contribute to the detection of void due to a higher penetration depth of the signals at these frequencies than at higher frequencies.

Electrical field intensity distribution for the proposed antenna and specimen 2 with and without void at frequencies of 3.74 and 6.95 GHz is shown in Fig. 5.20. It can be seen from Fig. 5.20 that there is significant disturbance of the electric field at the interface between free space (air) and surface of concrete slab due to a high relative (to free space) dielectric constant of concrete. Then, electromagnetic waves propagate in a uniform block (i.e. without contrast between the slab and the block,

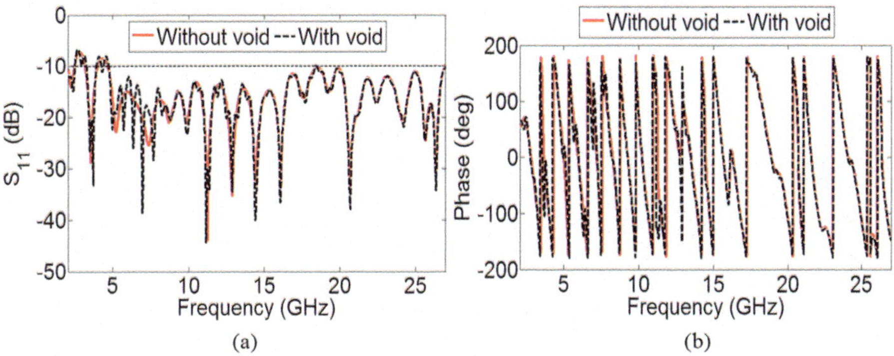

Fig. 5.19 Simulated (**a**) magnitude and (**b**) phase of the reflection coefficient with and without void inside specimen 2

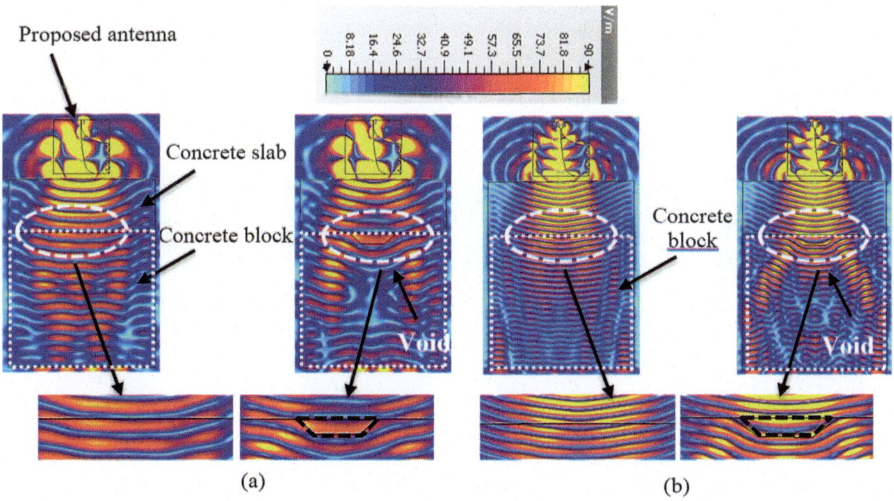

Fig. 5.20 Electrical field intensity distribution in the proposed antenna and specimen 2 (left) without and (right) with void at frequencies of (**a**) 3.74 GHz and (**b**) 6.95 GHz

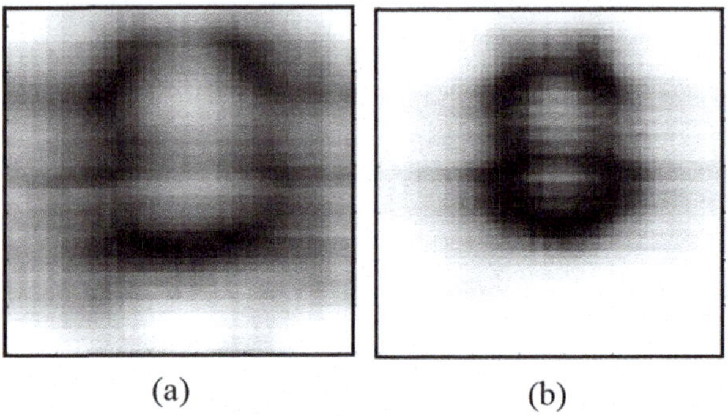

Fig. 5.21 Holographic slices obtained at distances of (**a**) 100 mm and (**b**) 112 mm from the top of the slab

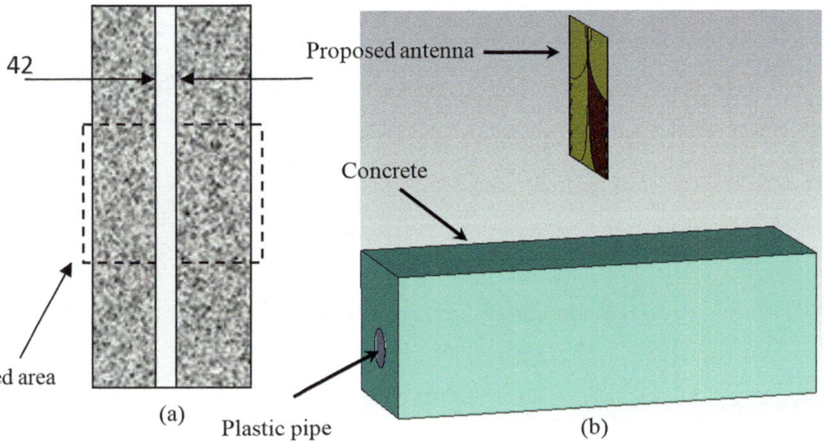

Fig. 5.22 (**a**) Schematic of a cross-sectional view and (**b**) model in CST of the proposed antenna placed at a distance of 60 mm from the centre of the concrete possessing plastic pipe

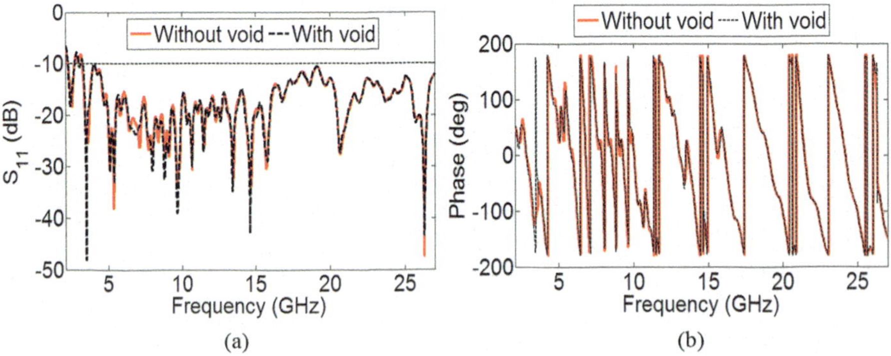

Fig. 5.23 Simulated (**a**) magnitude and (**b**) phase of the reflection coefficient with and without air-filled plastic pipe (void) embedded in the concrete

and void) with multiple reflections from its boundaries and splitting of the main lobe along their propagation. Figure 5.20 clearly shows significant changes of the electric field distribution at both frequencies when the void is present inside specimen 2. In addition, the comparison of field distributions at the two frequencies clearly indicates different values of depth of penetration in concrete at different frequencies; namely, penetration depth at higher frequency (i.e. at 6.95 GHz) is less than that at lower frequency (i.e. at 3.74 GHz). Therefore, it could be expected that thc signals at high frequencies would not sense the void in this specimen. The results of the field distribution explain the behaviour of the magnitude and phase of reflection coefficient over the entire frequency band shown in Fig. 5.19a, b. It should be noted that the cavity in both specimen 1 and specimen 2 had relatively sharp edges and their indications are prominent at the images of voids.

- **Imaging results for specimen 2**

Figure 5.21a, b shows two holographic slices obtained when the microwave signal was synthetically focusing at the interface between concrete block and slab (c.f. Fig. 5.21a), and the middle of the cavity (c.f. Fig. 5.21b), i.e. 100 and 112 mm from the top of the slab, respectively. It can be seen that Fig. 5.21a indicates the cavity which is slightly distorted by reflection/scattering of waves from edges resulting in the increased size of cavity indication compared to its actual size and non-uniform background of the image. On the other hand, when the microwave signal was focusing inside the cavity, the indication of the cavity is clearly visible in the image, background is uniform (due to absence of the cavity edges at sliced plane) and its size is very close to the actual size of the cavity (c.f. Fig. 5.21b).

Overall, the images of specimen 2 demonstrate a good capability of the proposed technique to penetrate inside concrete and to provide sufficient backscattered signal (signal-to-noise ratio) for the generation of appropriate images of its flaws such as void with a high-range resolution.

5.3.4 Proposed Antenna and Concrete with Plastic Pipe

To demonstrate the applicability of the proposed RTSAVA-B for the detection of void without sharp edges inside concrete, an air-filled cylindrical plastic pipe was embedded in a concrete beam (referred to as specimen 3) at the stage of its fabrication, i.e. during pouring of fresh concrete in a mould. Specimen 3 had dimensions of 100 mm by 100 mm by 400 mm and a 28 mm diameter plastic pipe was embedded in concrete at a distance of 60 mm from the top surface of specimen 3.

- **Simulation results for specimen 3**

To investigate the magnitude of reflection coefficient, S_{11} in dB, a model of specimen 3 possessing air-filled cylindrical plastic pipe embedded in concrete has been created in CST and simulation was performed. Figure 5.22 displays a model of the proposed antenna placed at a distance of 60 mm from the centre of the concrete possessing plastic pipe in CST.

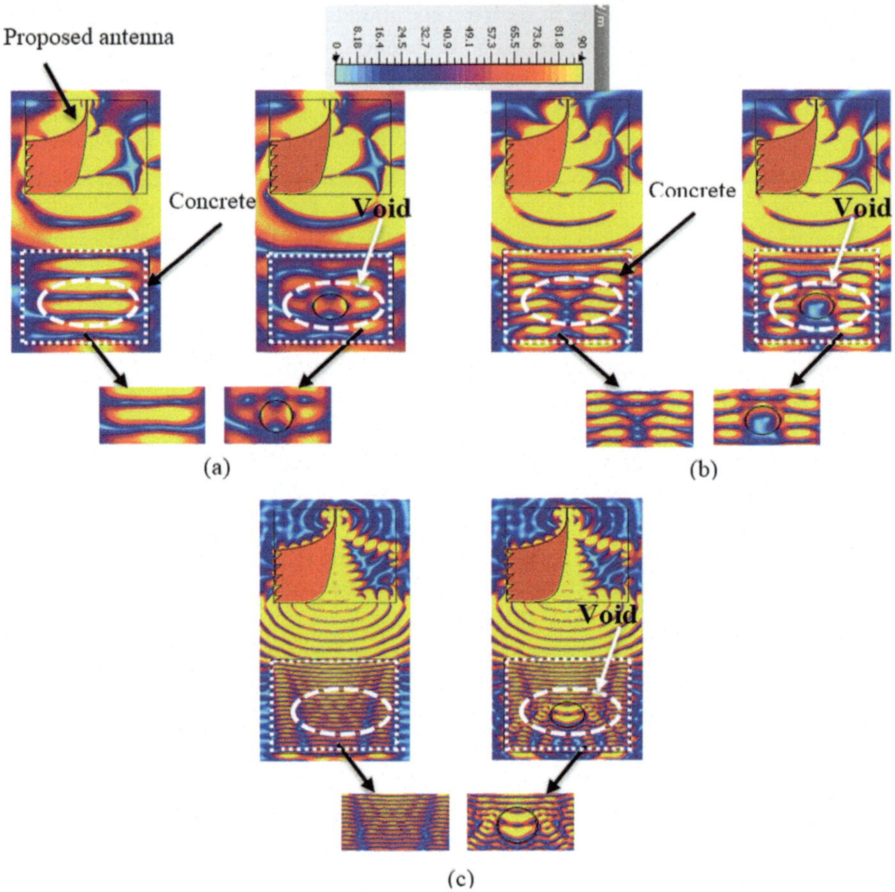

Fig. 5.24 Electrical field intensity distribution in the proposed antenna and specimen with and without void at frequencies of (**a**) 3.47 GHz, (**b**) 5.36 GHz and (**c**) 14.62 GHz

Figure 5.23a, b shows the magnitude and phase of reflection coefficient at a standoff distance of 60 mm for the dry concrete frequency range of 2–27 GHz. It can be noticed that there is a visible difference between values of the magnitude and values of the phase of reflection coefficient for the concrete with and without pipe/void at a frequency range from 2 to ~16 GHz.

Electrical field intensity distribution for the proposed antenna and specimen 3 with and without pipe/void at frequencies of 3.47, 5.36 and 14.62 GHz is shown in Fig. 5.24. A few observations can be made from Fig. 5.24 that are similar to those made for specimen 2 including a significant disturbance of the electric field at the interface between free space (air) and surface of concrete, multiple reflections inside concrete specimen and splitting of the main lobe. Figure 5.24 also shows changes of the electric field distribution at all three frequencies when the pipe/void is present inside specimen 3. Additional observations are as follows.

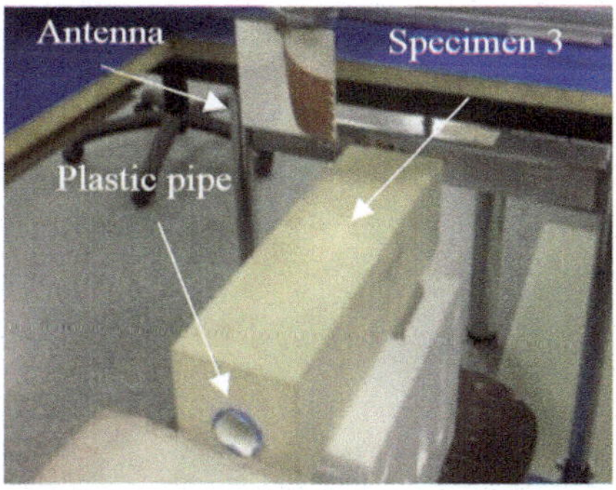

(a)

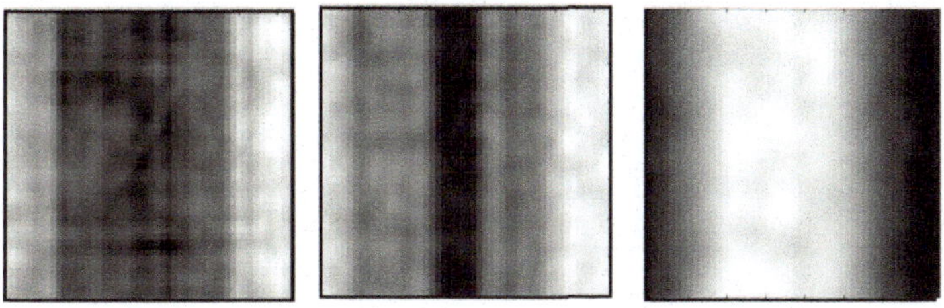

(b) (c) (d)

Fig. 5.25 (**a**) Picture of the antenna testing specimen 3 at a standoff distance of 60 mm and images when microwave signal was focused at the (**b**) top, (**c**) middle and (**d**) bottom of the specimen 3

The electromagnetic waves penetrated through specimen 3 and reflected from its bottom at all three frequencies. As a result, an indication of the bottom is visible in the slice shown in Fig. 5.25d. On the other hand, negligible changes of the magnitude and phase of the reflection coefficient caused by the presence of the pipe/void at higher frequencies can be attributed to relative small reflection of electromagnetic waves from smooth surface of the pipe/void which can be masked by strong reflections from the faces and edges of specimen 3.

- **Imaging results for specimen 3**

Figure 5.25a shows specimen 3 tested by the proposed antenna at a standoff distance of 60 mm using a scanned area of 150 mm by 150 mm and steps of 3 mm by 3 mm (c.f. Fig. 5.22a). Figure 5.25b–d shows horizontal slices of the 3D image when the microwave signal was focused at the plane of location of top, middle and bottom of specimen 3, respectively. These 2D images clearly show the top surface of specimen 3 (c.f. Fig. 5.25b), the air-filled pipe (c.f. Fig. 5.25c) and the bottom of specimen 3 (c. f. Fig. 5.25d).

5.4 Summary

In this chapter a technique including bent inner edges of the top and bottom radiators and regular triangular slits in saw-shaped configuration has been employed on conventional antipodal Vivaldi antenna to extend the low end of frequency band up to 2 GHz, and increase antenna gain at lower frequencies. This technique mitigates the dependence of the bandwidth requirement to antenna length and higher dielectric constant. In addition, the proposed antenna demonstrates improved F-to-B ratio and has low cross-polarisation of <-15 dB. The group time delay changes around 2.6 ns with a fluctuation of less than 0.1 ns, resulting in the antenna transmit and receive signals without serious distortion. To show the capability of the proposed antenna for ultra-wideband microwave imaging applications and to highlight penetration depth and range resolution of the imaging system with the antenna in the case of construction materials and structures, concrete specimens possessing voids as flaws or structural features such as an air-filled plastic pipe have been designed, fabricated and imaged. The results show that the proposed antenna along with the imaging system can be used for non-destructive testing and evaluation of concrete-based construction materials and structures.

References

Abbosh, A. M. (2009). Miniaturized microstrip-fed tapered-slot antenna with ultrawideband performance. *IEEE Antennas and Wireless Propagation Letters, 8*, 690–692.

Bai, J., Shi, S., & Prather, D. W. (2011). Modified compact antipodal Vivaldi antenna for 4–50-GHz UWB application. *IEEE Transactions on Microwave Theory and Techniques, 59*, 1051–1057.

Bourqui, J., Okoniewski, M., & Fear, E. C. (2010). Balanced antipodal Vivaldi antenna with dielectric director for near-field microwave imaging. *IEEE Transactions on Antennas and Propagation, 58*, 2318–2326.

De Oliveira, A. M., Perotoni, M. B., Kofuji, S. T., & Justo, J. F. (2015). A palm tree antipodal Vivaldi antenna with exponential slot edge for improved radiation pattern. *IEEE Antennas and Wireless Propagation Letters, 14*, 1334–1337.

Fei, P., Jiao, Y.-C., Hu, W., & Zhang, F.-S. (2011). A miniaturized antipodal Vivaldi antenna with improved radiation characteristics. *IEEE Antennas and Wireless Propagation Letters, 10*, 127–130.

Lee, D.-H., Yang, H.-Y., & Cho, Y.-K. (2012). Tapered slot antenna with band-notched function for ultrawideband radios. *IEEE Antennas and Wireless Propagation Letters, 11*, 682–685.

Molaei, A., Kaboli, M., Mirtaheri, S. A., & Abrishamian, M. S. (2014). Dielectric lens balanced antipodal Vivaldi antenna with low cross-polarisation for ultra-wideband applications. *IET Microwaves, Antennas and Propagation, 8*, 1137–1142.

Moosazadeh, M., & Kharkovsky, S. (2015a, October 4–7). *Design of ultra-wideband antipodal Vivaldi antenna for microwave imaging applications*. In 2015 IEEE International Conference on Ubiquitous Wireless Broadband (ICUWB), pp. 1–4.

Moosazadeh, M., & Kharkovsky, S. (2015b). Development of the antipodal Vivaldi antenna for detection of cracks inside concrete members. *Microwave and Optical Technology Letters, 57*, 1573–1578.

Moosazadeh, M., & Kharkovsky, S. (2016). A compact high-gain and front-to-back ratio elliptically tapered antipodal Vivaldi antenna with trapezoid-shaped dielectric lens. *IEEE Antennas and Wireless Propagation Letters, 15*, 552–555.

Moosazadeh, M., Kharkovsky, S., & Case, J. T. (2016a). Microwave and millimetre wave antipodal Vivaldi antenna with trapezoid-shaped dielectric lens for imaging of construction materials. *IET Microwaves, Antennas and Propagation, 10*, 301–309.

Nassar, I. T., & Weller, T. M. (2015). A novel method for improving antipodal Vivaldi antenna performance. *IEEE Transactions on Antennas and Propagation, 63*, 3321–3324.

Natarajan, R., George, J. V., Kanagasabai, M., & Shrivastav, A. K. (2015). A compact antipodal Vivaldi antenna for UWB applications. *IEEE Antennas and Wireless Propagation Letters, 14*, 1557–1560.

Pandey, G., Verma, H., & Meshram, M. (2015). Compact antipodal Vivaldi antenna for UWB applications. *Electronics Letters, 51*, 308–310.

Schaubert, D. H., Kollberg, E. L., Korzeniowski, T. L., Thungren, T., Johansson, J., & Yngvesson, K. S. (1985). Endfire tapered slot antennas on dielectric substrates. *IEEE Transactions on Antennas and Propagation, 33*, 1392–1400.

Shaari, A., Millard, S., & Bungey, J. (2004). Modelling the propagation of a radar signal through concrete as a low-pass filter. *NDT and E International, 37*, 237–242.

Teni, G., Zhang, N., Qiu, J., & Zhang, P. (2013). Research on a novel miniaturized antipodal Vivaldi antenna with improved radiation. *IEEE Antennas and Wireless Propagation Letters, 12*, 417–420.

Wang, S., Chen, X., & Parini, C. (2007). *Analysis of ultra wideband antipodal Vivaldi antenna design*. In 2007 Loughborough Antennas and Propagation Conference. IEEE, pp. 129–132.

Wang, Z., Yin, Y., Wu, J., & Lian, R. (2016). A miniaturized CPW-fed antipodal Vivaldi antenna with enhanced radiation performance for wideband applications. *IEEE Antennas and Wireless Propagation Letters, 15*, 16–19.

Wu, J., Zhao, Z., Nie, Z., & Liu, Q.-H. (2014). A printed UWB Vivaldi antenna using stepped connection structure between slot-line and tapered patches. *IEEE Antennas and Wireless Propagation Letters, 13*, 698–701.

Zhu, F., Gao, S., Ho, A. T., Abd-Alhameed, R. A., See, C. H., Li, J., & Xu, J. (2012). Miniaturized tapered slot antenna with signal rejection in 5–6-GHz band using a balun. *IEEE Antennas and Wireless Propagation Letters, 11*, 507–510.

Chapter 6
Miniaturised Antipodal Vivaldi Antenna and Its Application for Detection of Void Inside Concrete Specimens

6.1 Introduction

In practice, the detection of flaws in construction materials such as concrete requires antennas operating at relatively low frequencies [e.g. 1 GHz (Millard et al. 2002)] to provide appropriate penetration depth of electromagnetic waves and wide-frequency bandwidth. To achieve this using conventional approaches, the antenna size will be large; therefore, miniaturisation technique (Zhu et al. 2012) may need to be applied to antenna to meet small size of antenna with lower operating frequency.

As stated in previous chapters, AVA has been used for microwave imaging applications. Various approaches have been reported in the literature for AVA miniaturisation. In Abbosh (2009), the top and bottom radiators of the antenna have been corrugated to reduce antenna size while keeping antenna performance constant. In Teni et al. (2013), regular slots on the antenna edges cause the antenna operate at a lower operating frequency with a smaller size. In Fei et al. (2011), a miniaturised AVA is presented wherein tapered slots on the antenna edges have been employed to lower the low end of the frequency band. An AVA with reduced size is reported in Karahan and Armagan Sahinkaya (2014) and Zhang et al. (2012). Recently, elliptically shaped tapered strips have been used to extend impedance bandwidth (Wang et al. 2016). However, impedance bandwidth of the referenced antennas is still limited and their achieved gain is still low at the low and/or high ends of the frequency band. It should be noted that the degradation of gain could be considered as the limits of bandwidth.

AVAs with dielectric lenses have been presented in Chaps. 3 and 4; however, their impedance bandwidths are restricted at lower operating frequencies (i.e. 5 GHz in Chap. 3 and 3.4 GHz in Chap. 4) and their gains are low at lower frequencies (e.g. <6 dB at 5 GHz). As a result, these antennas can only be used for imaging low-loss construction materials such as plasterboard sheets (Moosazadeh et al. 2016). An

M. Moosazadeh, *Antipodal Vivaldi Antennas for Microwave Imaging of Construction Materials and Structures*,
https://doi.org/10.1007/978-3-030-05566-0_6

AVA with regular triangular shaped slit has been presented in Chap. 5 to lower the low end of frequency band; however, their impedance bandwidths are restricted at lower operating frequencies (i.e. 2 GHz in Chap. 2).

In this chapter, a miniaturised AVA (MAVA) with a minimum operating frequency of 1 GHz and relatively high gain for the entire frequency band is proposed to be applied for microwave imaging of high-loss as well as low-loss construction materials at a frequency range from 1 to 30 GHz. The models of the proposed antenna and concrete beam possessing foam mimicking void created in CST have been presented. A prototype of the antenna is fabricated and employed as part of microwave imaging system to detect voids inside concrete beam.

The chapter includes three main sections. Modification process and improvement of antenna gain for the MAVA are documented in the second section. The performance of the designed antennas in free space has been provided in the third section. Section 6.4 presents the results of application of the proposed antenna for detection of void inside the concrete beam.

6.2 Modification Process for Miniaturised Antenna

A step-by-step modification process has been employed to design and optimise the performance of the proposed antenna. First, a CAVA is designed as a reference. Second, the CAVA is modified to create a small AVA (SAVA) with reduced size. Third, to extend the low end of the frequency band and further reduce the size of the antenna, the inner edges of the top and bottom radiators of the SAVA have been bent, creating an MAVA. To enhance gain at lower frequencies, regular slit edge (RSE) is designed, resulting in a miniaturised regular slit-edge antipodal Vivaldi antenna (MRSEAVA). Finally, a half-elliptical shaped dielectric lens (HEDL) as an extension of the antenna substrate is added to MRSEAVA and a novel antenna referred to as MRSEAVA-HEDL is designed to feature high gain and F-to-B ratio.

6.2.1 Miniaturisation of Antipodal Vivaldi Antenna

As stated, to achieve the desired penetration depth, ṻFF;s, for high-loss construction materials such as concrete, lower operating frequency is required. For this purpose, the lowest operating frequency targeted in this study is 1 GHz (Millard et al. 2002). A detailed design of the antenna follows. Figure 6.1a–c demonstrates the configuration of three intermediate antennas showing the step-by-step modification process for the following designed antennas: CAVA, SAVA and MAVA. The length, L_{SUBN}, and width of the aperture, W_{AN}, of the designed antennas can be obtained from the following equations (Schaubert et al. 1985):

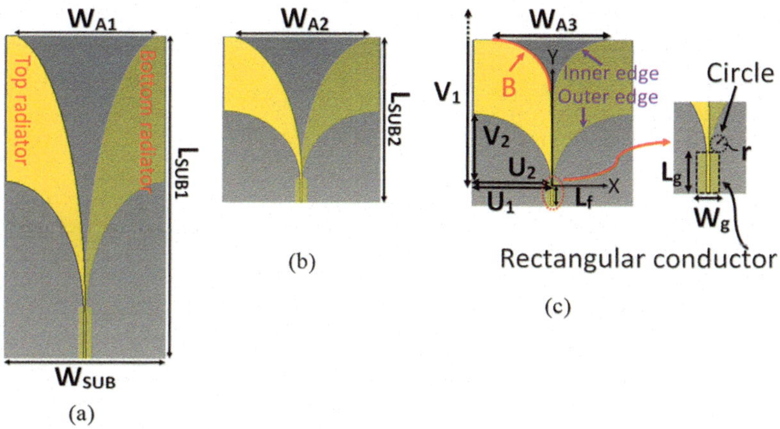

Fig. 6.1 Configuration of the designed AVAs in *XY*-plane showing the modification process: (**a**) CAVA, (**b**) SAVA and (**c**) MAVA and zoom-in of the ground plane configuration

$$L_{\text{SUBN}} = \frac{c}{f_{\text{L}}}\sqrt{\frac{2}{\varepsilon_r + 1}} \tag{6.1}$$

$$W_{\text{AN}} = \frac{c}{2f_{\text{L}}\sqrt{\varepsilon_r}} \tag{6.2}$$

where c, N, f_{L} and ε_r are speed of light in free space, number corresponding to the step in the modification process, lower working frequency and dielectric constant, respectively, as shown in Fig. 6.1. All above-mentioned antennas were printed on a Rogers RO4003C substrate with thickness of 0.508 mm, dielectric constant of 3.38 and loss tangent of 0.0027. The antennas are fed through a microstrip line with a width of 1.14 mm in order to match with the 50-Ω coaxial line. The top and bottom radiators are formed by the intersection of two ellipses for only one quadrant with major radii (U_1 and U_2) and the secondary radii (V_1 and V_2) as shown in Fig. 6.1c. The tapered ground plane section is created from the intersection of a rectangular shaped conductor with dimension of $W_g \times L_g$ and two anti-faced quarter circles with radius r.

The simulated magnitude of reflection coefficient (S_{11}) in dB of the designed antennas is plotted in Fig. 6.2. According to Eqs. (6.1) and (6.2) with $f_{\text{L}} = 1$ GHz, the CAVA should have L_{SUB1} of 202 mm and W_{A1} of 81.74 mm. It is observed that the CAVA has a bandwidth from 5 to 30 GHz as shown in Fig. 6.2a and a poor impedance matching occurs between 1 and 5 GHz as shown in Fig. 6.2b. On the other hand, the CAVA size is big, which is a problem since a small antenna size is desired so that the antenna may be integrated into microwave imaging systems. In the next step of the antenna design, length of the antenna is shortened as L_{SUB2} of 100 mm (i.e. Fig. 6.1b) (i.e. SAVA). It can be seen that this step has bandwidth from 4.5 to

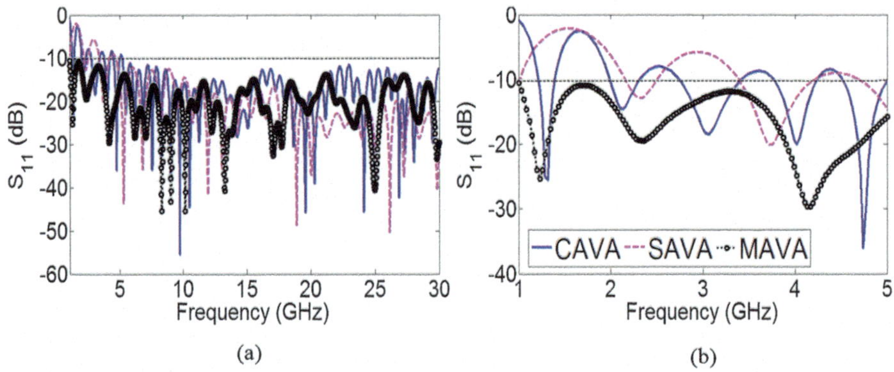

Fig. 6.2 Simulated magnitude of the reflection coefficient (S_{11}) in dB for the designed antennas showed in Fig. 6.1 at (**a**) 1–30 GHz and (**b**) 1–5 GHz

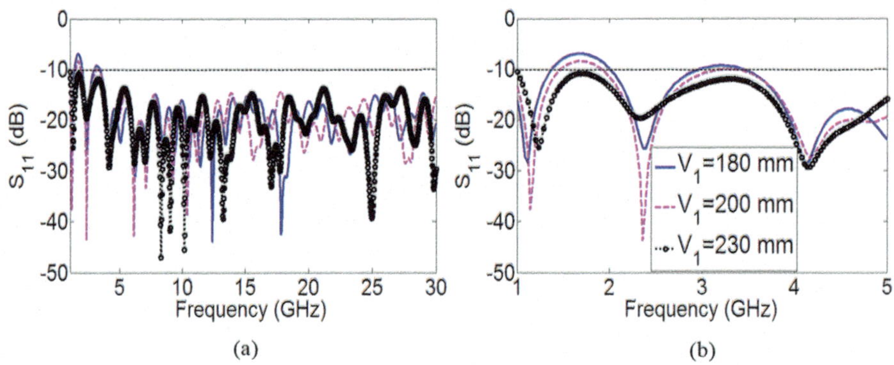

Fig. 6.3 Simulated magnitude of the reflection coefficient (S_{11}) in dB of the designed MAVA with $B = 35$ for various values of V_1 at bandwidths of (**a**) 1–30 GHz and (**b**) 1–5 GHz

30 GHz as shown in Fig. 6.2a and a poor impedance matching occurs between 1 and 4.5 GHz as shown in Fig. 6.2b. To extend the low end of frequency band of the antenna in the next step of the design, the inner edges of the top and bottom radiators have been properly bent with B in Fig. 6.1c where B is the opening elliptical bending rate. As shown in Fig. 6.2a, the designed MAVA has bandwidth from 1 to 30 GHz and improves impedance matching between 1 and 5 GHz as shown in Fig. 6.2b. In summary, the antenna length of MAVA states 50% of the length CAVA with lowest operating frequency of 1 GHz.

In the MAVA, the opening elliptical rate, B, and length of radiator, V_1, play a main role; therefore it is essential that the B and V_1 should be optimised under condition of obtaining the widest impedance bandwidth and antenna miniaturisation by parametric study. To do that, first, by keeping B constant, S_{11} has been optimised by changing V_1. Figure 6.3a, b shows the magnitude of reflection coefficient of MAVA with $B = 35$ and various values of V_1 at whole and lower frequencies bands,

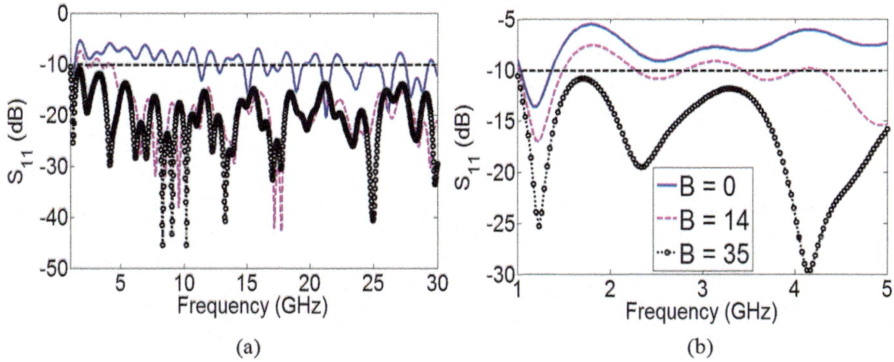

Fig. 6.4 Simulated magnitude of the reflection coefficient (S_{11}) in dB of the designed MAVA with $V_1 = 230$ mm for various values of B at bandwidths of (**a**) 1–30 GHz and (**b**) 1–5 GHz

respectively, indicating that the impedance matching has been improved at lower frequencies by increasing V_1 up to 230 mm.

Second, while maintaining V_1 at 230 mm, we studied variation of B on impedance function. Figure 6.4a, b shows the magnitude of reflection coefficient of the MAVA with $V_1 = 230$ mm and various values of B at whole and lower frequencies bands, respectively, indicating that the magnitude of S_{11} decreases as B increases from 0 to 35 at lower frequencies.

Therefore, the MAVA with the design parameters B of 35 and V_1 of 230 mm provides an impedance bandwidth from 1 to 30 GHz and a miniaturised geometry. The physical mechanism of the designed MAVA with performance of below cut-off frequency of the SAVA (i.e. 5 GHz) can be elaborated as follows. The low end of frequency band is defined by the width of the antenna aperture which can be calculated by Eq. (6.2). As width of antenna aperture increases, the wave's energy separates and provides radiated space waves. In the proposed MAVA, current path is extended along the inner edges of the top and bottom layers with a suitable value of B, leading to extending of lower end of frequency band.

In order to further understand the effect of bending of the inner edges of the antenna at low frequencies, E-field distribution at 1.5 GHz and total radiation efficiency are shown in Fig. 6.5a, b, respectively. It is clear that the intensity of E-field in radiation region 'A' is weak and it can be attributed to high reflection of waves in the antenna leading to high magnitude of the reflection coefficient of the SAVA as shown in Fig. 6.2b. As a result low radiation efficiency of the antenna can be observed in Fig. 6.5b. By bending inner edges of the top and bottom radiators, significant E-field concentration is achieved in radiation region 'C', resulting in lower magnitude of the reflection coefficient, and higher radiation efficiency of the MAVA as shown in Figs. 6.2b and 6.5b, respectively.

Gain of the proposed MAVA is investigated as well. The gain variation at the entire frequency range (1–30 GHz) and at lower frequencies (1–5 GHz) for the selected B values is plotted in Fig. 6.6a, b, respectively. As shown, the highest gain corresponds to a B of 35. However, the MAVA gain is low (i.e. <0 dB) at 1 GHz.

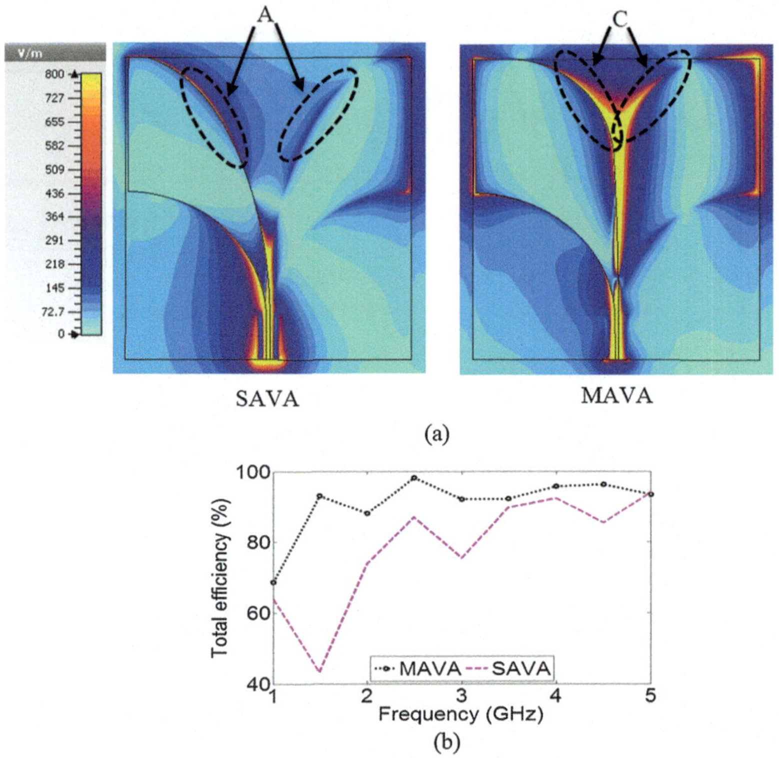

Fig. 6.5 (**a**) E-field distribution in *XY*-plane of the designed SAVA and MAVA at 1.5 GHz and (**b**) total efficiency of the antennas at a bandwidth of 1–5 GHz

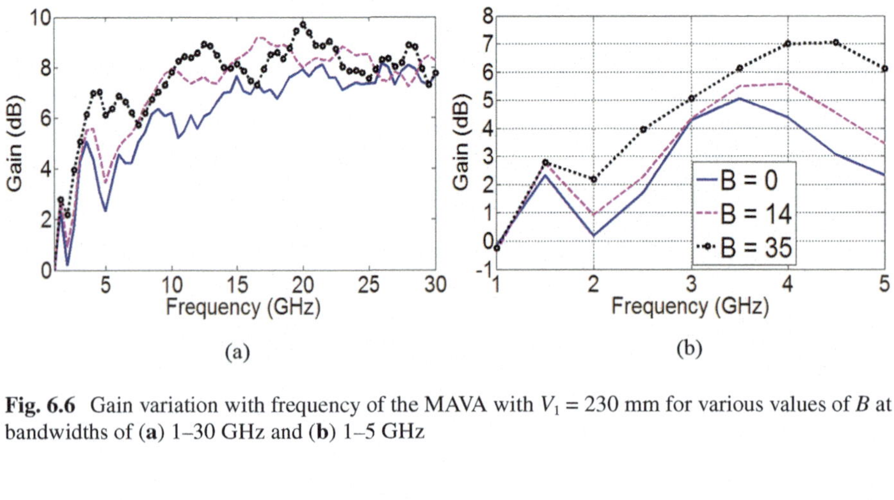

Fig. 6.6 Gain variation with frequency of the MAVA with $V_1 = 230$ mm for various values of B at bandwidths of (**a**) 1–30 GHz and (**b**) 1–5 GHz

6.2.2 *Improvement of Antenna Gain*

As displayed in earlier section, the MAVA gain is low at a frequency of 1 GHz. To enhance gain at 1 GHz, RSE has been implemented to the designed MAVA (i.e. MRSEAVA) as shown in Fig. 6.7a. The gain variation versus frequency for both the MAVA and MRSEAVA at the whole (1–30 GHz) and lower (1–4 GHz) frequency bands is shown in Fig. 6.7b, c. As depicted from Fig. 6.7c, gain is increased by RSE at 1 GHz (i.e. >0 dB).

However, gain of the MAVA and MRSEAVA at middle and higher frequencies is low as shown in Fig. 6.7b. To increase antenna gain, an extension of antenna substrate as HEDL is also implemented to the designed MRSEAVA (i.e. MRSEAVA-HEDL) as shown in Fig. 6.8a. A parametric study on the various radius values of extension, R, for gain variation is performed as shown in Fig. 6.8b, resulting in the highest gain obtained at $R \geq 40$ mm.

The simulated electrical field intensity distribution on XY-plane for the designed antennas at frequencies of 20 and 30 GHz is displayed in Fig. 6.9, indicating that the endfire radiation statues are apparent. As expected, the proposed MRSEAVA-HEDL focused electromagnetic waves along Y-direction.

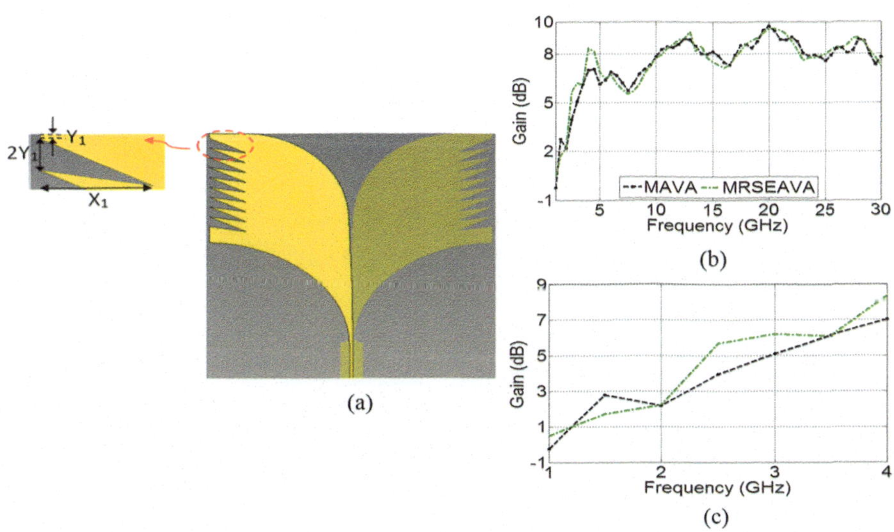

Fig. 6.7 (**a**) Configuration of the designed MRSEAVA and gain of the MAVA and MRSEAVA at (**b**) 1–30 GHz and (**c**) 1–4 GHz

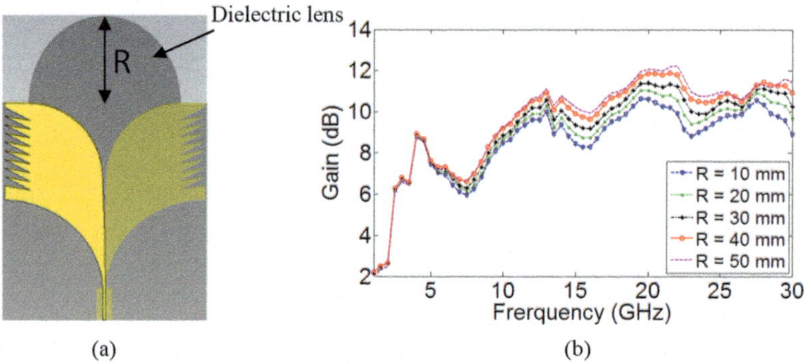

Fig. 6.8 (**a**) Configuration of the designed MRSEAVA-HEDL and (**b**) variation of the gain with frequency for the various elliptical lengths of *R*

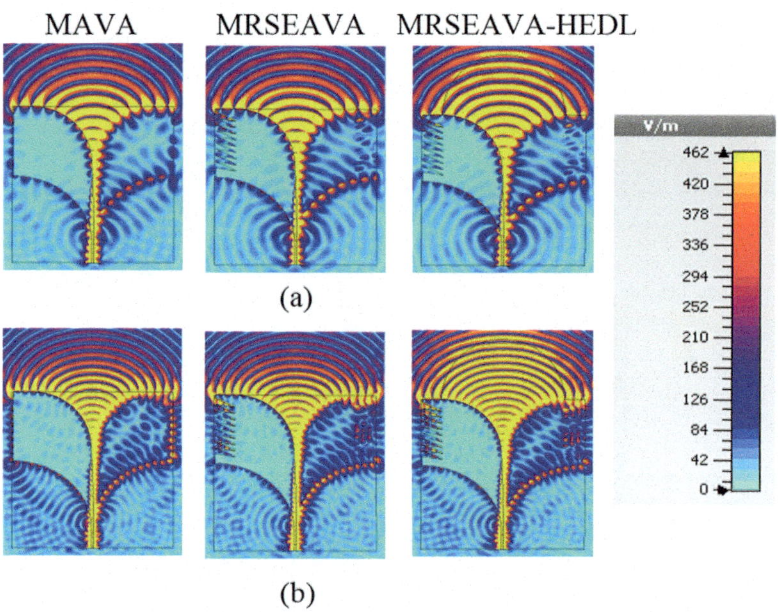

Fig. 6.9 Simulated E-field distribution on *XY*-plane with normalised linear scale of the colour map for the designed antennas at (**a**) 20 GHz and (**b**) 30 GHz

6.3 Performance of the Designed Antennas in Free Space

As a result of a detailed parametric study on different parameters, the optimal parameters of the designed MRSEAVA-HEDL are as follows: W_{SUB} = 96 mm, L_{SUB2} = 100 mm, W_{A3} = 71.46 mm, $U_1 = (W_{SUB} + W_f)/2$, $V_1 = 4.73 \times U_1$,

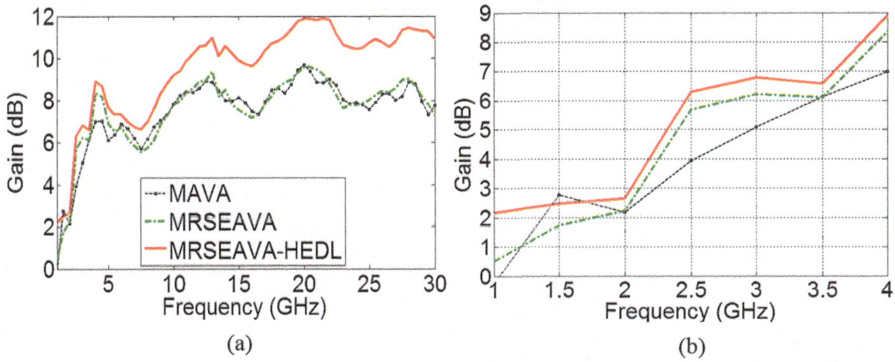

Fig. 6.10 Gain variation of the designed antennas in endfire direction at (**a**) 1–30 GHz and (**b**) lower frequencies, 1–4 GHz

$U_2 = (W_{SUB} - W_f)/2$, $V_2 = 0.9 \times U_2$, $B = 35$, $X_1 = 12$ mm, $Y_1 = 2$ mm, $R = 40$ mm, $r = (W_g - W_f)/2$, $L_f = 13$ mm, $L_g = 17.5$ mm and $W_g = 7.5$ mm.

The gain variation of the MAVA, MRSEAVA and MRSEAVA-HEDL at the entire frequency range (1–30 GHz) and at lower frequencies (1–4 GHz) is depicted in Fig. 6.10a, b, respectively. It can be found that the HEDL not only enhances antenna gain at higher frequencies but also improves gain at lower frequencies as shown in Fig. 6.10b. In summary, highest gain is related to MRSEAVA-HEDL, resulting in antenna gain reaching to its maximum value of 12 dB at 20 GHz as shown in Fig. 6.10a.

To specify the worthiness of the proposed antenna, its performance is compared with the referenced antennas in accordance with the dimensions and dielectric constant as shown in Table 6.1 (λ_0 is the free-space wavelength corresponding to the lowest operating frequency). The obtained bandwidth with small size of antenna in De Oliveira et al. (2015) and Abbosh (2009) is at the cost of high dielectric constant. Though the antenna in Wang et al. (2016) has also relatively small size, the proposed antenna has wider bandwidth.

The normalised *E*-plane (*XY*-plane) and *H*-plane (*YZ*-plane) radiation patterns of the designed MAVA and MRSEAVA-HEDL at the different frequencies are displayed in Fig. 6.11. According to the indicated radiation patterns for both *E*- and *H*-planes, the MRSEAVA-HEDL demonstrates good directional radiation patterns in spite of miniaturised size. Through investigating Fig. 6.11, it is observable that pattern beamwidths for *H*-plane are wider at frequencies of below 10 GHz and then narrow as frequencies increase to above 16 GHz.

It can be clearly also seen that the MRSEAVA-HEDL shows an improvement of F-to-B ratio when compared with MAVA. In order to effectively show improvement of the F-to-B ratio, in Fig. 6.12 is plotted F-to-B ratio variation versus frequency for the designed MAVA and MRSEAVA-HEDL. As shown in Fig. 6.12, a noticeable improvement of the F-to-B ratio for the MRSEAVA-HEDL is obtained compared to the MAVA. The maximum F-to-B ratio (39.45 dB) is achieved with the MRSEAVA-

Table 6.1 The specification and performances of the referenced Vivaldi antennas and proposed antenna

References	Dimensions (mm^3)	B.W. (GHz)	E_r	Gain (dB) at Lowest frequency	Highest frequency
Chamaani et al. (2011)	86.2 × 94 × 0.76	2.6–11	3.5	6	9
Lee et al. (2012)	50 × 50 × 0.8	2.4–11.2	4.4	2.5	2.5
Wu et al. (2014)	41 × 48 × 0.8	3–15.1	4.4	5.5	6
Bai et al. (2011)	64 × 61 × 0.524	4–50	2.2	2.5	12.5
In et al. (2012)	52 × 145 × 0.508	3.1–10.6	2.2	3.5	12.5
Natarajan et al. (2015)	42 × 36 × 1.6	3.7–18	4.4	1.8	5
Herzi et al. (2013)	90 × 86 × 0.76	2.2–15	3.5	–	–
Ashraf et al. (2015b)	120 × 160 × 1.57	0.8–18	2.2	5	9
De Oliveira et al. (2015)	36.3 × 59.8 × 0.6	4.73–11	6.15	5.5	8
Wang et al. (2012)	50.6 × 101 × 1.5	6–18	3.38	8.8	11.3
Molaei et al. (2014)	50 × 166 × 3.15	3–18	2.55	2.8	12
Moosazadeh and Kharkovsky (2015b)	96 × 100 × 0.508	2–27	3.38	2.8	11.2
Nassar and Weller (2015)	66 × 140 × 1.5	2–32	2.94	−4	−1
Abbosh (2009)	30 × 25 × 0.64	1.8–10.8	10.2	–	7.8
Fei et al. (2011)	48 × 60 × 1	2.4–14	4.6	3.7	7.5
Cao et al. (2014)	44 × 50 × 3	2.1–10	2.65	1	8
Zhang et al. (2012)	48 × 60 × 1	2.77–12	4.4	6.8	10.5
Zhu et al. (2012)	38 × 39 × 0.813	3.05–10.3	3.38	2	2
Teni et al. (2013)	50 × 66.4 × 1	4–30	4.5	5	5.5
Pandey et al. (2015)	50 × 56 × 0.8	2–12	4.4	1.5	4.5
Wang et al. (2016)	90 × 93.5 × 0.8	1.32–17	2.65	6.8	9
Moosazadeh and Kharkovsky (2016)	40 × 90 × 0.508	3.4–40	3.38	8	15
Huang and Hsu (2011)	78.7 × 91 × 0.5	3-18	3.38	5.5	10.2
Chu et al. (2015)	100 × 96 × 1.6	2.3–15	4.4	5.5	5
Proposed antenna, MRSEAVA-HEDL	96 × 140 × 0.508	1–30	3.38	2.2	11

HEDL at a frequency of 16 GHz which is approximately more than twice that for the MAVA (i.e. 16.86 dB).

To estimate time signal distortion, the group time delay behaviour of the proposed MRSEAVA-HEDL is investigated. To do this, two profiles of the MRSEAVA-HEDL are put in endfire direction (*Y*-direction in Fig. 6.1) with a distance of 150 mm to act as transmitting (*TX*) and receiving (*RX*) antennas as shown in Fig. 6.13 (inset). It can be clearly seen that the supreme group time delay of the proposed antenna is around 1.6 ns with fluctuation of less than 0.1 ns in most frequencies. It can be found that the proposed antenna is able to transmit and receive signals without serious distortion.

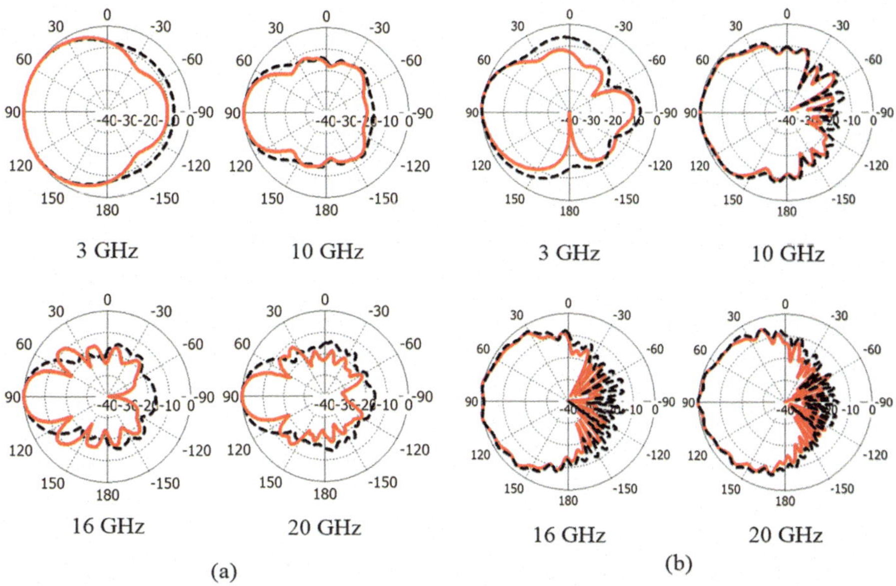

Fig. 6.11 The normalised (**a**) *H*- and (**b**) *E*-plane radiation patterns of the MAVA (dashed line) and MRSEAVA-HEDL (solid line) at different frequencies

Fig. 6.12 Variation of the F-to-B ratio with frequency for the designed MAVA and MRSEAVA-HEDL

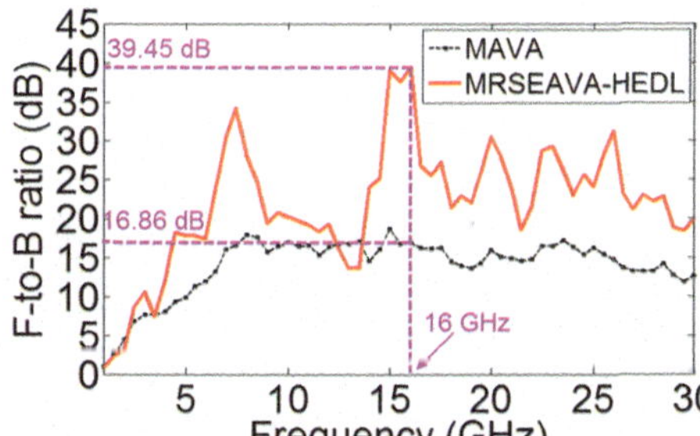

Fig. 6.13 Group delay of the proposed MRSEAVA-HEDL at a bandwidth 1–30 GHz and two-antenna arrangement facing each other with 150 mm separation in endfire direction (inset).

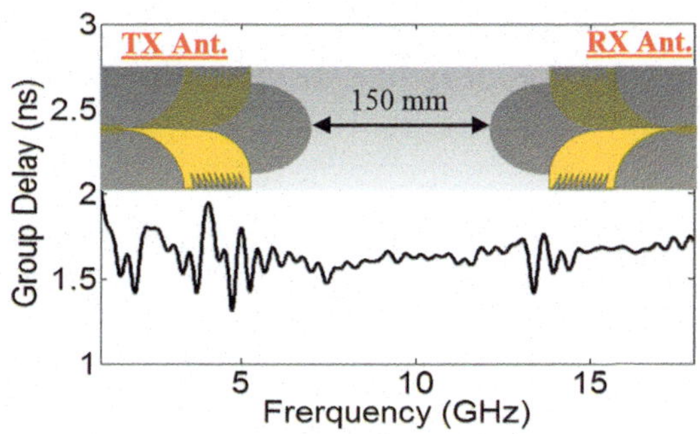

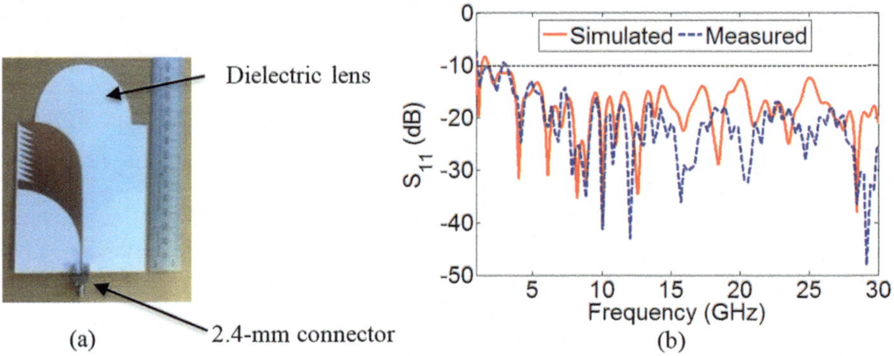

Fig. 6.14 (**a**) Photograph of the fabricated MRSEAVA-HEDL (top view) and (**b**) simulated and measured magnitude of the S_{11} of the proposed antenna

6.4 Microwave Imaging with the Fabricated MRSEAVA-HEDL

6.4.1 Fabricated Antenna and Measurement Imaging Approach

A prototype of the MRSEAVA-HEDL was fabricated and tested with PNA to validate simulated S_{11}, and apply for microwave imaging of construction materials. A photo of the fabricated antenna with 2.4 mm connector is shown in Fig. 6.14a. The measured and simulated S_{11} of the proposed antenna is plotted in Fig. 6.14b, indicating that the antenna has a bandwidth from 1 to 30 GHz for S_{11} <−10 dB. There are some discrepancies for measured result which can be due to inaccuracy in fabrication of antenna and connector.

To demonstrate the effectiveness of the proposed antenna for wideband microwave imaging and to highlight the obtainable penetration depth and range resolution, concrete specimens possessing voids were fabricated and imaged.

6.4.2 Imaging of Concrete Beam Possessing Foam Mimicking Void

A 3-month-old concrete beam with dimensions of 400 × 100 × 100 mm^3 possessing two foam inserts has been fabricated to simulate voids with dimensions of ~25 × 32 × 20 mm^3 as shown in Fig. 6.15a, b. It should be noted that the localisation of inserts had a small shift (~5 mm) from the centre of the beam resulting in different distances between each foam insert and sides of the specimen; that is, they were

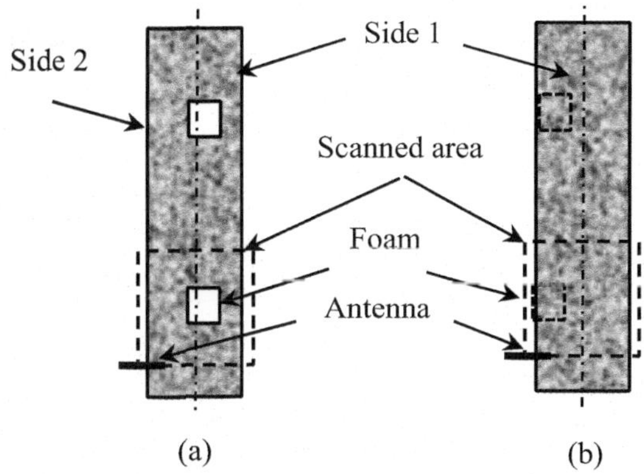

Fig. 6.15 Schematic of concrete beam with two foam inserts tested by the proposed antenna: (**a**) top and (**b**) side views

closer to Side 1 (~28 mm) than to Side 2 (~38 mm) as shown in Fig. 6.15a. The concrete beam was scanned with a 3 mm step size along each axis over the scanned area of 105 mm by 105 mm as shown in Fig. 6.15a, b.

6.4.2.1 Irradiation of the Top Surface of the Specimen

- **Simulation results**

The model of the proposed antenna placed at the standoff distance of *d* from above the centre of the concrete beam (dielectric constant of 4.5 and loss tangent of 0.0125) possessing foam at top surface is created in CST as shown in Fig. 6.16.

Figure 6.17 describes simulated magnitude and phase of the reflection coefficient over the entire frequency range (1–30 GHz) for case of concrete beam and foam insert at the top surface when the antenna was located above the centre of the concrete beam at a standoff distance of 10 mm. It can be clearly found from Fig. 6.17 that there is difference between values of the magnitude as well as values of phase of reflection coefficient for the concrete beam with and without void at a frequency range from 5.6 to 30 GHz and this difference increases when the frequency increases. It means that signals with frequencies within frequency range from 5.6 to 30 GHz provide the main contribution to the detection of foam (void) placing at the top surface.

To further understand this effect, electrical field intensity distribution for the proposed antenna and concrete beam with and without foam (void) at frequencies of 5.6–10.2 GHz is shown with colour bar in Fig. 6.18. The foam (void) designated with dashed-line sections in Fig. 6.18 demonstrates the influence of void in concrete on the field distribution and shows changes of the electric field distribution when the foam is present at the top surface of concrete beam. Figure 6.18 demonstrates the significant disturbance of the electric field at the interface between frees space (air) and concrete beam, in particular at a frequency of 10.2 GHz. As shown, electromag-

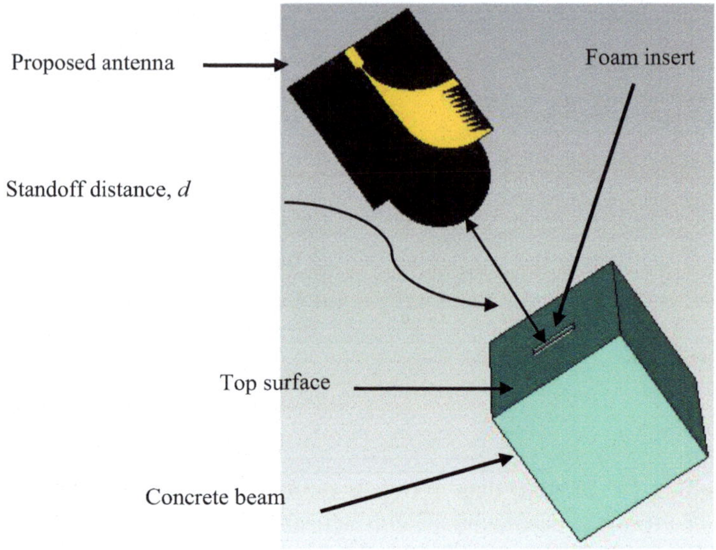

Fig. 6.16 Model of the proposed antenna placed at the standoff distance of *d* from above the centre of the concrete beam possessing foam at the top surface in CST

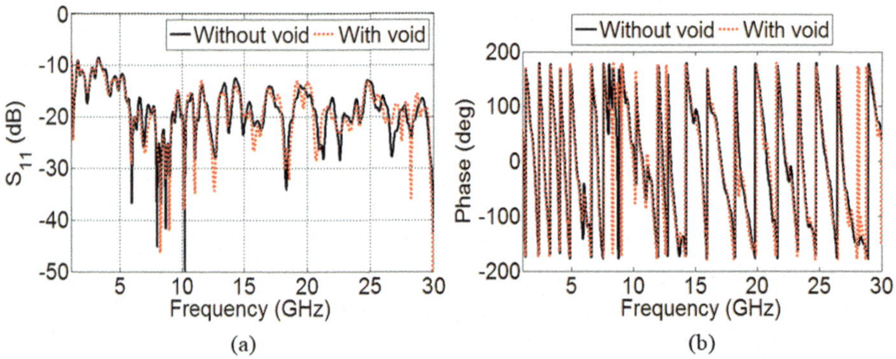

Fig. 6.17 Simulated (**a**) magnitude and (**b**) phase of the reflection coefficient with and without foam (void) inserted at the top surface in the concrete beam at a standoff distance of 10 mm

netic waves propagate in concrete beam with multiple reflections from its boundaries and splitting of the main lobe [i.e. Fig. 6.18b with foam (void)] along their propagation and as a result an indication of the void is visible in the slice shown in Fig. 6.21a. The results of the field distribution explain the behaviour of the magnitude and phase of reflection coefficient over the entire frequency band shown in Fig. 6.17.

Figure 6.19 describes simulated magnitude and phase of the reflection coefficient over the entire frequency range (1–30 GHz) for case of concrete beam and

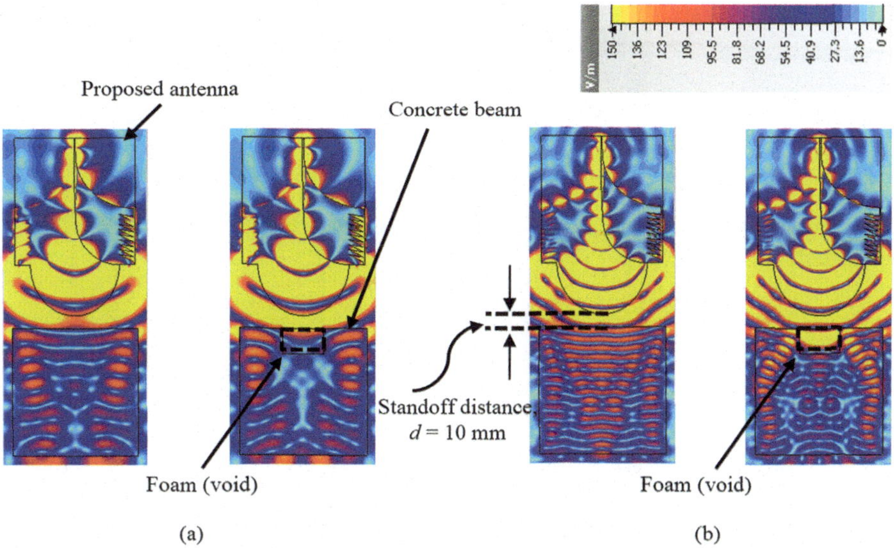

Fig. 6.18 Electrical field intensity distribution in the proposed antenna and concrete beam without and with foam (void) at frequencies of (**a**) 5.6 GHz and (**b**) 10.2 GHz at a standoff distance of 10 mm

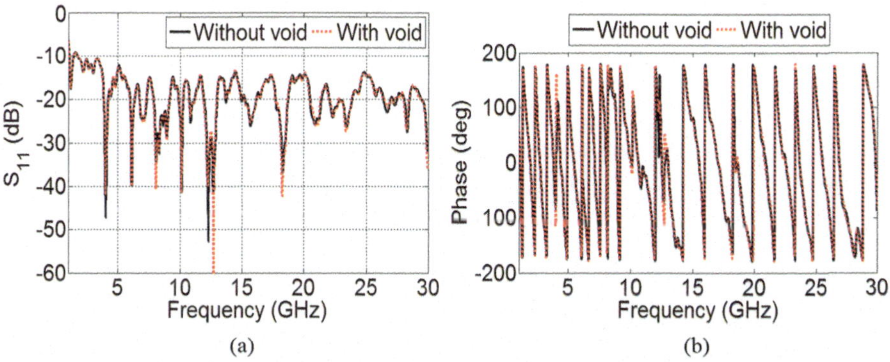

Fig. 6.19 Simulated (**a**) magnitude and (**b**) phase of the reflection coefficient with and without foam (void) inserted at the top surface in the concrete beam at a standoff distance of 100 mm

foam insert at the top surface when the antenna was located above the centre of the concrete beam at a standoff distance of 100 mm. It can be clearly found from Fig. 6.19 that there is visible difference between values of the magnitude and values of phase of reflection coefficient for the concrete with and without foam (void) at a frequency range from 1 to 20 GHz. It means that signals with frequencies within a frequency range from 1 to 20 GHz provide the main contribution to the detection of

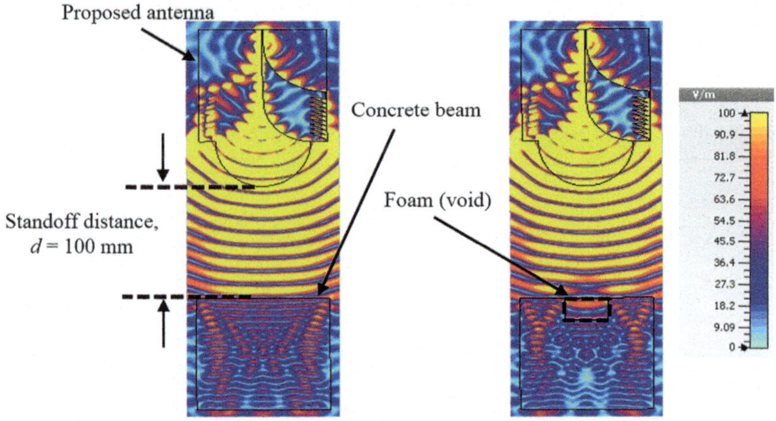

Fig. 6.20 Electrical field intensity distribution in the proposed antenna and concrete beam without and with foam (void) at a frequency of 12.74 GHz at a standoff distance of 100 mm

void due to a higher penetration depth of the signals at these frequencies than at higher frequencies.

To further understand this effect, electrical field intensity distribution for the proposed antenna and concrete beam with and without foam (void) at a frequency of 12.74 GHz is shown with colour bar in Fig. 6.20. The foam (void) designated with dashed-line sections in Fig. 6.20 demonstrates influence of void in concrete on the field distribution and shows changes of the electric field distribution when the foam is present at the top surface of concrete beam. Figure 6.20 demonstrates the significant disturbance of the electric field at the interface between frees space (air) and concrete beam. As shown, electromagnetic waves propagate in concrete beam with multiple reflections from its boundaries and splitting of the main lobe along their propagation and as a result an indication of the void is visible in the slice shown in Fig. 6.21b. The results of the field distribution explain the behaviour of the magnitude and phase of reflection coefficient over the entire frequency band shown in Fig. 6.19.

- **Imaging results**

To evaluate the obtainable range resolution at different standoff distances, the top surface of specimen with an exposed foam insert was imaged. Figure 6.21a, b shows holographic slices of the 3D images obtained at a standoff distance of 10 and 100 mm, respectively. Figure 6.21a, b shows the holographic slices corresponding to from left to right of the columns of the top surface of foam insert and 9 and 15 mm below that. Rectangular indications of the foam insert are clearly visible for slices corresponding to the top and bottom of the insert (i.e. left and right images in Fig. 6.21a). This can be attributed to the contribution of a strong reflection of the signal from the edges of concrete similar to reflection from the edges of specimen

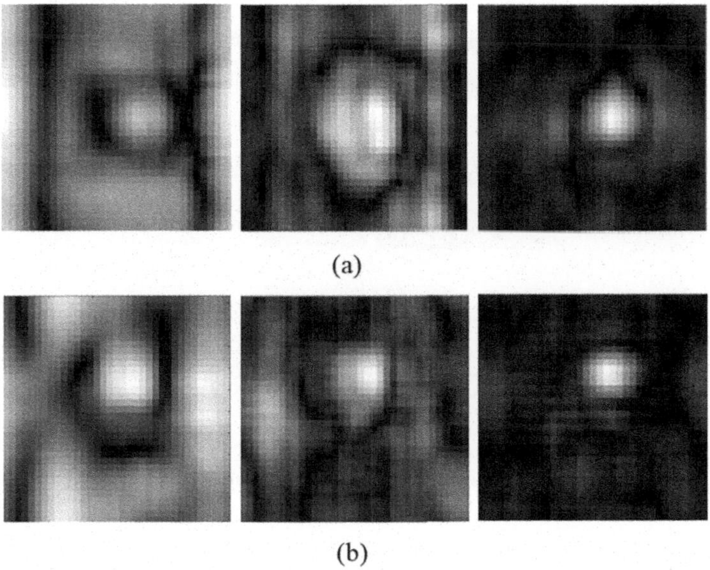

Fig. 6.21 Holographic slices of 3D images obtained at standoff distances of (**a**) 10 mm and (**b**) 100 mm from the top of specimen

as shown on the left image in Fig. 6.21a. A non-rectangular indication of foam insert can be seen in the central image in Fig. 6.21a where there is no contribution of concrete edges. Indications of the foam insert obtained at a standoff distance of 100 mm are shown in Fig. 6.21b. As expected, they are slightly weaker than those obtained at a standoff distance of 10 mm but they are still clearly visible. The result shows that range resolution of a few millimetres in concrete has been achieved at a wide range of standoff distances in this investigation.

6.4.2.2 Irradiation of the Side of the Specimen

- **Simulation results**

The model of the proposed antenna placed at a distance of 10 mm alongside the concrete beam possessing foam is created in CST as depicted in Fig. 6.22. Figure 6.23 shows simulated magnitude and phase of the reflection coefficient for case of concrete beam and foam insert over the entire frequency range (1–30 GHz) when the antenna was located alongside the concrete beam at a standoff distance of 10 mm. It can be clearly seen from Fig. 6.23 that there is difference between values of the magnitude and values of phase of reflection coefficient for the concrete with and without foam/void at higher frequencies, e.g. 22.2 GHz.

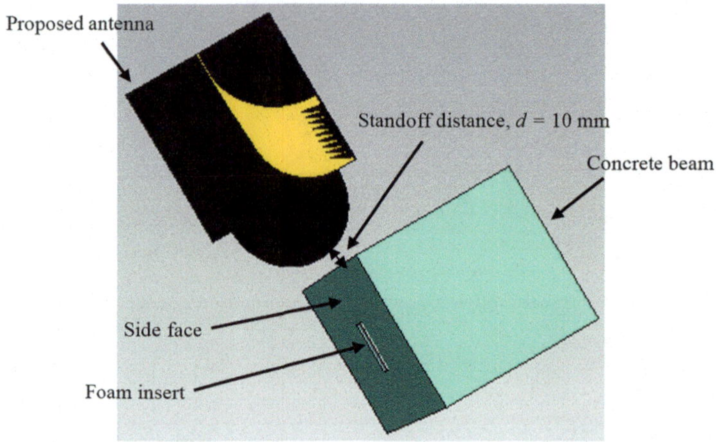

Fig. 6.22 Model of the proposed antenna placed at the distance of 10 mm alongside the concrete beam possessing foam in CST

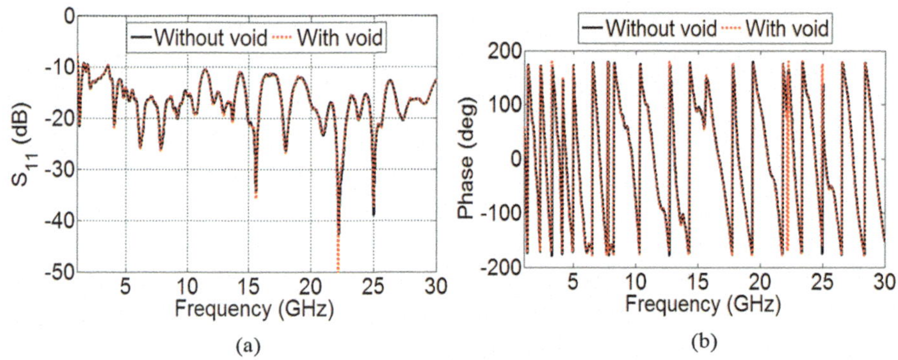

Fig. 6.23 Simulated (**a**) magnitude and (**b**) phase of the reflection coefficient with and without foam (void) inserted in the concrete beam at a standoff distance of 10 mm

Electrical field intensity distribution for the proposed antenna and concrete beam with and without foam (void) insert at a frequency of 22.2 GHz is shown in Fig. 6.24. A few observations can be made from Fig. 6.24 that are similar to those made in Fig. 6.18 including a significant disturbance of the electric field at the interface between free space (air) and surface of concrete beam, multiple reflections inside concrete beam and splitting of the main lobe. Figure 6.24 also shows changes of the electric field distribution at a frequency of 22.2 GHz when the foam/void is present inside concrete beam as designated with dashed lines. The electromagnetic waves penetrated through concrete beam and reflected along its side face. As a result, an indication of the void is visible in the slice shown in Fig. 6.25b.

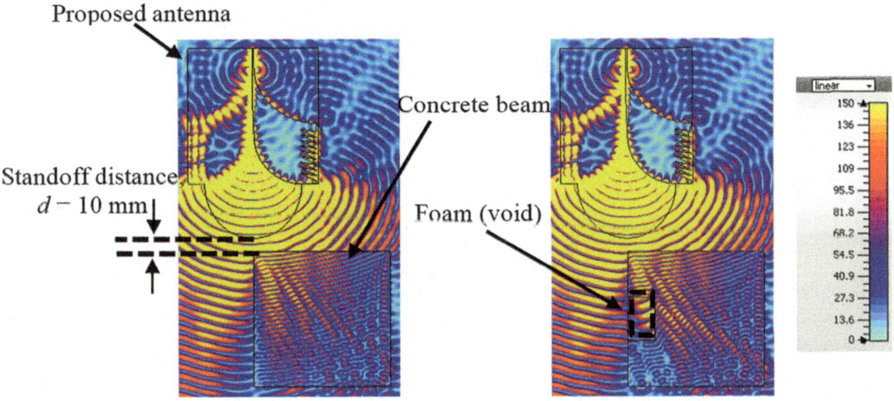

Fig. 6.24 Electrical field intensity distribution (left) without and (right) with foam (void) at a standoff distance of 10 mm when antenna is placed alongside the concrete beam at a frequency of 22.2 GHz

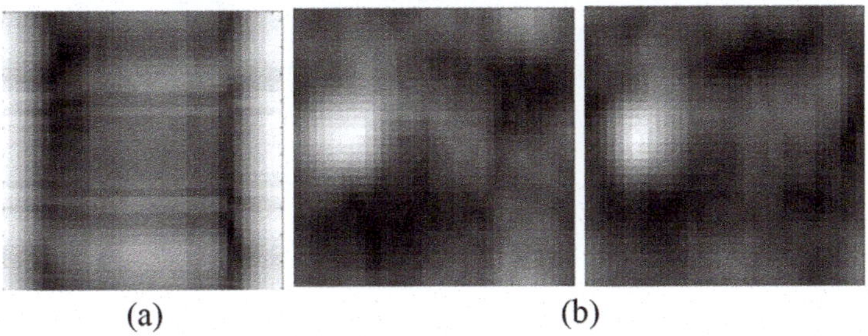

Fig. 6.25 Holographic slices of 3D image obtained at a standoff distance of 10 mm from Side 1 of the specimen when microwave signal was focused at (**a**) the surface of Side 1 and (**b**) the foam insert

- **Imaging results**

To estimate the depth of penetration of microwave signals transmitted and received by the imaging system with the proposed antenna, different sides of the specimen were imaged. As mentioned, the localisation of the foam inserts is non-symmetrical with respect to the different sides of specimen.

Figure 6.25 shows holographic slices of the 3D image of specimen obtained when Side 1 was imaged. The image of the surface of Side 1 is slightly distorted while indication of the foam insert is clearly visible as shown in Fig. 6.25a, b, respectively. When the side surface of the foam insert (left image in Fig. 6.25b) is imaged, its indication is slightly wider than that obtained when the microwave signal was focused 3 mm below the surface of foam. Again, it may be a result of a relatively strong reflection from the edge of concrete (right image in Fig. 6.25b). An indication of the foam insert was not obtained when Side 2 was imaged by the proposed antenna. It should be noted that the 3-month-old concrete specimen was not quite dry resulting in an increased attenuation in the specimen and decreased depth of penetration. Overall, ~60 mm round-trip propagation distance has been achieved in this investigation.

6.5 Summary

Minimum operating frequency of 1 GHz was achieved in the proposed miniaturised antenna by bending the inner edges of the top and bottom radiators of a conventional antipodal Vivaldi antenna. To increase gain at lower and higher frequencies, regular slit edge and a half-elliptical shaped dielectric lens as extension of antenna substrate have been applied, respectively. To show the capability of the proposed antenna and to highlight the penetration depth and range resolution of the imaging system with the proposed antenna, a concrete specimen possessing void has been fabricated and imaged. The proposed antenna provides better images of high-loss materials such as concrete than our previously designed antennas. The results show that the proposed antenna along with the imaging system can be used for non-destructive testing and evaluation of numerous construction materials including concrete materials and structures.

References

Abbosh, A. M. (2009). Miniaturized microstrip-fed tapered-slot antenna with ultrawideband performance. *IEEE Antennas and Wireless Propagation Letters, 8*, 690–692.

Ashraf, M. A., Jamil, K., Sebak, A., Shoaib, M., Alhekail, Z., Alkanhal, M., & Alshebeili, S. (2015b). Modified antipodal Vivaldi antenna with shaped elliptical corrugation for 1–18 GHz UWB Application. *Applied Computational Electromagnetics Society Journal, 30*(1), 68–77.

Bai, J., Shi, S., & Prather, D. W. (2011). Modified compact antipodal Vivaldi antenna for 4–50-GHz UWB application. *IEEE Transactions on Microwave Theory and Techniques, 59*, 1051–1057.

Cao, Y., Lei, J., Wei, Y., & Zhu, L. (2014). *A compact BAVA design with corrugated edge*. In 2014 3rd Asia-Pacific Conference on Antennas and Propagation (APCAP). IEEE, pp. 259–262.

Chamaani, S., Mirtaheri, S. A., & Abrishamian, M. S. (2011). Improvement of time and frequency domain performance of antipodal Vivaldi antenna using multi-objective particle swarm optimization. *IEEE Transactions on Antennas and Propagation, 59*, 1738–1742.

Chu, H. B., Shirai, H., & Dao, C. N. (2015). *Effect of curvature of antipodal structure on Vivaldi antennas*. In 2015 IEEE International Symposium on Antennas and Propagation & USNC/URSI National Radio Science Meeting. IEEE, pp. 2331–2332.

De Oliveira, A. M., Perotoni, M. B., Kofuji, S. T., & Justo, J. F. (2015). A palm tree antipodal Vivaldi antenna with exponential slot edge for improved radiation pattern. *IEEE Antennas and Wireless Propagation Letters, 14*, 1334–1337.

Fei, P., Jiao, Y.-C., Hu, W., & Zhang, F.-S. (2011). A miniaturized antipodal Vivaldi antenna with improved radiation characteristics. *IEEE Antennas and Wireless Propagation Letters, 10*, 127–130.

Herzi, R., Gharbi, R., Zairi, H., & Gharsallah, A. (2013, March 18–21). *A tuneable antipodal Vivaldi antenna for UWB application*. In 2013 10th International Multi Conference on Systems, Signals & Devices (SSD), pp. 1–4.

Huang, T.-J. & Hsu, H.-T. (2011). *Antipodal dual exponentially tapered slot antennas Antipodal dual exponentially tapered slot antennas (DETSA) with corrugations for front-to-back ratio improvement*. In 2011 IEEE International Workshop on Electromagnetics, Applications and Student Innovation (iWEM). IEEE, pp. 48–51.

In, D. M., Lee, M. J., Kim, D., Oh, C. Y., & Kim, Y. S. (2012). Antipodal linearly tapered slot antenna using unequal half-circular defected sides for gain improvements. *Microwave and Optical Technology Letters, 54*, 1963–1965.

Karahan, M., & Armagan Sahinkaya, D. S. (2014). *A reduced size antipodal Vivaldi antenna design for wideband applications*. In 2014 IEEE Antennas and Propagation Society International Symposium (APSURSI). IEEE, pp. 1588–1589.

Lee, D.-H., Yang, H.-Y., & Cho, Y.-K. (2012). Tapered slot antenna with band-notched function for ultrawideband radios. *IEEE Antennas and Wireless Propagation Letters, 11*, 682–685.

Millard, S., Shaari, A., & Bungey, J. (2002). Field pattern characteristics of GPR antennas. *NDT and E International, 35*, 473–482.

Molaei, A., Kaboli, M., Mirtaheri, S. A., & Abrishamian, M. S. (2014). Dielectric lens balanced antipodal Vivaldi antenna with low cross-polarisation for ultra-wideband applications. *IET Microwaves, Antennas and Propagation, 8*, 1137–1142.

Moosazadeh, M., & Kharkovsky, S. (2015b). Development of the antipodal Vivaldi antenna for detection of cracks inside concrete members. *Microwave and Optical Technology Letters, 57*, 1573–1578.

Moosazadeh, M., & Kharkovsky, S. (2016). A compact high-gain and front-to-back ratio elliptically tapered antipodal Vivaldi antenna with trapezoid-shaped dielectric lens. *IEEE Antennas and Wireless Propagation Letters, 15*, 552–555.

Moosazadeh, M., Kharkovsky, S., & Case, J.T. (2016). Microwave and millimetre wave antipodal Vivaldi antenna with trapezoid-shaped dielectric lens for imaging of construction materials. *IET Microwaves, Antennas and Propagation, 10*(3), 301–309.

Nassar, I. T., & Weller, T. M. (2015). A novel method for improving antipodal Vivaldi antenna performance. *IEEE Transactions on Antennas and Propagation, 63*, 3321–3324.

Natarajan, R., George, J. V., Kanagasabai, M., & Shrivastav, A. K. (2015). A compact antipodal Vivaldi antenna for UWB applications. *IEEE Antennas and Wireless Propagation Letters, 14*, 1557–1560.

Pandey, G., Verma, H., & Meshram, M. (2015). Compact antipodal Vivaldi antenna for UWB applications. *Electronics Letters, 51*, 308–310.

Schaubert, D. H., Kollberg, E. L., Korzeniowski, T. L., Thungren, T., Johansson, J., & Yngvesson, K. S. (1985). Endfire tapered slot antennas on dielectric substrates. *IEEE Transactions on Antennas and Propagation, 33*, 1392–1400.

Teni, G., Zhang, N., Qiu, J., & Zhang, P. (2013). Research on a novel miniaturized antipodal Vivaldi antenna with improved radiation. *IEEE Antennas and Wireless Propagation Letters, 12*, 417–420.

Wang, P., Zhang, H., Wen, G., & Sun, Y. (2012). Design of modified 6-18 GHz balanced antipodal Vivaldi antenna. *Progress in Electromagnetics Research C, 25*, 271–285.

Wang, Z., Yin, Y., Wu, J., & Lian, R. (2016). A miniaturized CPW-fed antipodal Vivaldi antenna with enhanced radiation performance for wideband applications. *IEEE Antennas and Wireless Propagation Letters, 15*, 16–19.

Wu, J., Zhao, Z., Nie, Z., & Liu, Q.-H. (2014). A printed UWB Vivaldi antenna using stepped connection structure between slot-line and tapered patches. *IEEE Antennas and Wireless Propagation Letters, 13*, 698–701.

Zhang, Q., Fu, L., Shi, S., Xu, J., & Wei, G. (2012). *A miniaturized fermi TSA with improved radiation characteristics For UWB Application*. In 2012 8th International Symposium on Communication Systems, Networks & Digital Signal Processing (CSNDSP). IEEE, pp. 1–4.

Zhu, F., Gao, S., Ho, A. T., Abd-Alhameed, R. A., See, C. H., Li, J., & Xu, J. (2012). Miniaturized tapered slot antenna with signal rejection in 5–6-GHz band using a balun. *IEEE Antennas and Wireless Propagation Letters, 11*, 507–510.

Chapter 7
Comb-Shaped Slit Antipodal Vivaldi Antenna and Its Application for Detection of Void Inside Concrete Specimens

7.1 Introduction

In Chap. 6, an AVA has been presented to extend the low end of the frequency band to 1 GHz. However, its gain is low at low frequencies (2.2 dB at 1 GHz and 2.6 dB at 2 GHz) and this performance can be critical in some applications.

To obtain high gain at low frequencies, an UWB elliptically tapered AVA is developed for detection of flaws in high-loss materials such as concrete. First, description of the specimen and models of the proposed antenna and concrete beam possessing void created in CST has been presented. Then, a prototype of the antenna is fabricated and employed as part of the microwave imaging system to verify simulation results and to detect voids inside the concrete beam. The capability of the proposed antenna for UWB imaging for the purpose of the detection of voids inside the concrete beam is demonstrated. The results of microwave imaging of voids inside concrete at different standoff distances between the proposed antenna and the surface of concrete are presented.

The chapter includes three main sections. In the second section, the configuration of the designed antennas is presented. The third section presents antenna performance in free space, while Sect. 7.4 demonstrates the applicability of the proposed antenna for detection of void inside a concrete beam.

7.2 Configuration of Designed Antennas

A detailed design of the antenna is as follows. The configuration of the designed antennas, namely CAVA and AVA with bending inner edges of top and bottom radiators (AVA-B) and rectangular slits in periodic (comb-shaped) configuration on AVA-B (CSAVA-B), is shown in Fig. 7.1a–c. Dimensions of the substrate

M. Moosazadeh, *Antipodal Vivaldi Antennas for Microwave Imaging of Construction Materials and Structures*,
https://doi.org/10.1007/978-3-030-05566-0_7

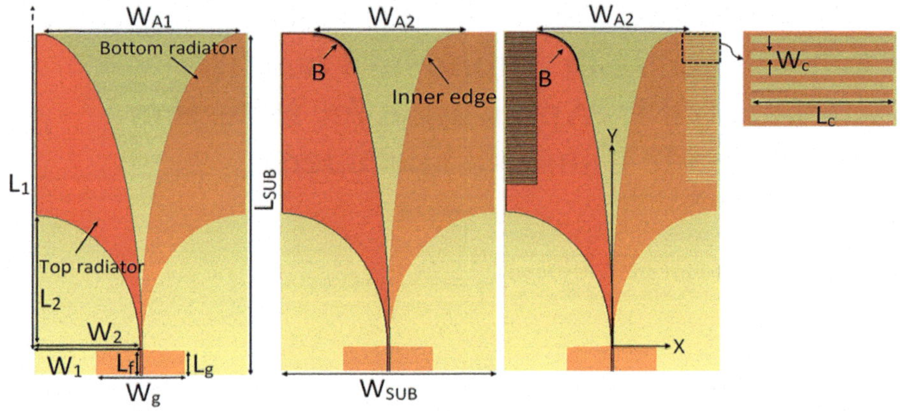

Fig. 7.1 Schematic of the designed antennas in *XY*-plane: (**a**) CAVA, (**b**) AVA-B and (**c**) CSAVA-B (Moosazadeh et al. 2017)

($W_{SUB} \times L_{SUB}$) of the designed antennas are set to be 120 × 202 mm^2. The designed antennas are printed on Rogers RO4003C with a relative dielectric constant of 3.38, a loss tangent of 0.0027 and a thickness of 0.508 mm. We have used a thin substrate made of materials with a relatively low dielectric constant to avoid the degradation of radiation characteristics at higher frequencies since increasing electrical thickness of the substrate causes unwanted modes.

According to previous research, the effective substrate thickness should meet the requirement of $0.005\lambda_0 < h_{eff} < 0.03\lambda_0$, where λ_0 is the wavelength at the centre frequency of the operating bandwidth and h_{eff} is $h(\sqrt{\varepsilon_r} - 1)$. The antenna is fed by a microstrip line with a width (W_f) of 1.14 mm in order to match the 50-Ω coaxial connector. For this purpose the 3D (*XYZ*) simulation mesh matrix of 657 × 1026 × 22 is utilised with around 14 million mesh cells. The top and bottom radiators are formed by the intersection of two ellipses for only one quadrant with major radii (W_1 and W_2) and the secondary radii (L_1 and L_2) as shown in Fig. 7.1a. The ground plane is elliptically tapered from the intersection of a rectangular shaped conductor with the dimension of $W_g \times L_g$ and two anti-faced quarter ellipses with the major radii of $(W_g - W_f)/2$ and the secondary radii of 1 mm.

Figure 7.2a shows the simulated magnitude of the reflection coefficient (S_{11}) in dB of the designed antennas. To better show the efficacy of comb-shaped slits on S_{11} at lower frequencies, the frequency band of 1–5 GHz is selected and illustrated in Fig. 7.2b. It is clearly observed that the cut-off frequency for CAVA is 4 GHz and a poor impedance matching occurs for 1–4 GHz as shown in Fig. 7.2b. In order to lower the cut-off frequency, inner edges of the radiator have been bent with *B* as shown in Fig. 7.1b where *B* is the opening elliptical bending rate. Figure 7.2b shows that impedance matching is improved within 1.8–4 GHz except at 2.4 GHz. To extend the low end of frequency band of the AVA-B, in the next step of the design

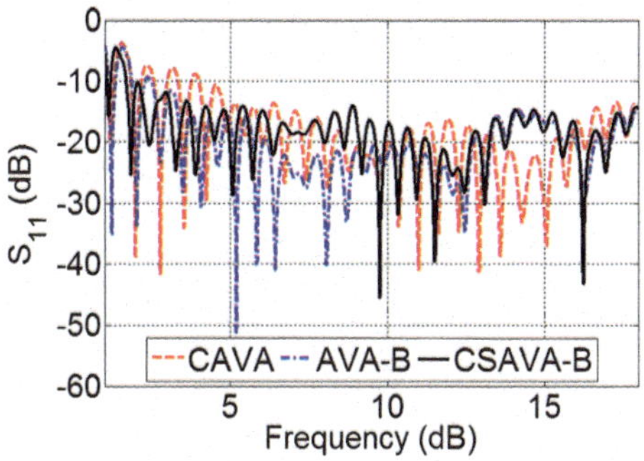

(a)

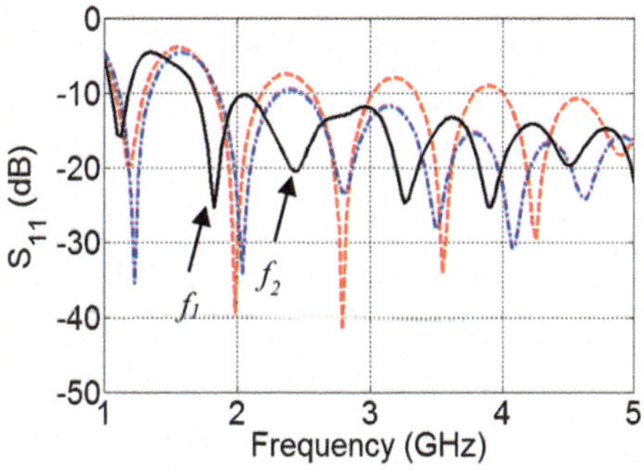

(b)

Fig. 7.2 Simulated magnitude of the reflection coefficient (S_{11}) in dB for the designed antennas showed in Fig. 7.1 at (**a**) bandwidth of 1–18 GHz and (**b**) zoom-in at low frequencies of 1–5 GHz (Moosazadeh et al. 2017)

the comb-shaped slits have been employed as shown in Fig. 7.1c. It can be clearly seen from Fig. 7.2a that the designed CSAVA-B has a bandwidth from 1.65 to 18 GHz. The length of the employed slits L_c labelled in Fig. 7.1c is set to be approximately $0.25\lambda/\sqrt{\varepsilon_r}$, where λ is the wavelength corresponding to 2.4 GHz. It should be mentioned that width of the slits, W_c, does not impact the impedance bandwidth.

In order to better show the influence of the inner edge bending and comb-shaped slits on the antenna performance, surface current distribution and total radiation efficiency at a frequency band from 1.65 to 5 GHz are shown in Fig. 7.3a, b, respectively. As shown in Fig. 7.3a, the surface current is concentrated on the area of the edges of the radiators designated with 'A', leading to the high magnitude of the reflection coefficient of the CAVA. Employing rectangular slits in periodic (comb-shaped) configuration causes the surface currents to concentrate in region 'C'. As a result, the low end of frequency band is extended. Figure 7.3b shows the simulated total radiation efficiency versus frequency for CAVA and CSAVA-B at low frequencies from 1.65 to 5 GHz. As shown in Fig. 7.3b, antenna efficiency increases with comb-shaped slits and bending the inner edges of the top and bottom radiators. As a result, gain increases as is shown in Fig. 7.6.

Radiation patterns for the designed CAVA and CSAVA-B are investigated as well. The simulated *H*- and *E*-plane radiation patterns at the resonant frequencies f_1 and f_2 are shown in Fig. 7.4. It is clearly seen from Fig. 7.4 that lower side lobe levels and improved F-to-B ratio are achieved for the proposed CSAVA-B compared with the CAVA.

The optimised parameters of the designed CSAVA-B are as follows: $W_{A2} = 84$ mm, $W_1 = (W_{SUB} + W_f)/2$, $W_2 = (W_{SUB} - W_f)/2$, $L_1 = 229$ mm, $L_2 = 80$ mm, $W_f = 1.14$ mm, $L_f = 15$ mm, $W_g = 50$ mm, $L_g = 15$ mm, $W_c = 1$ mm, $L_c = 17$ mm and $B = 25$.

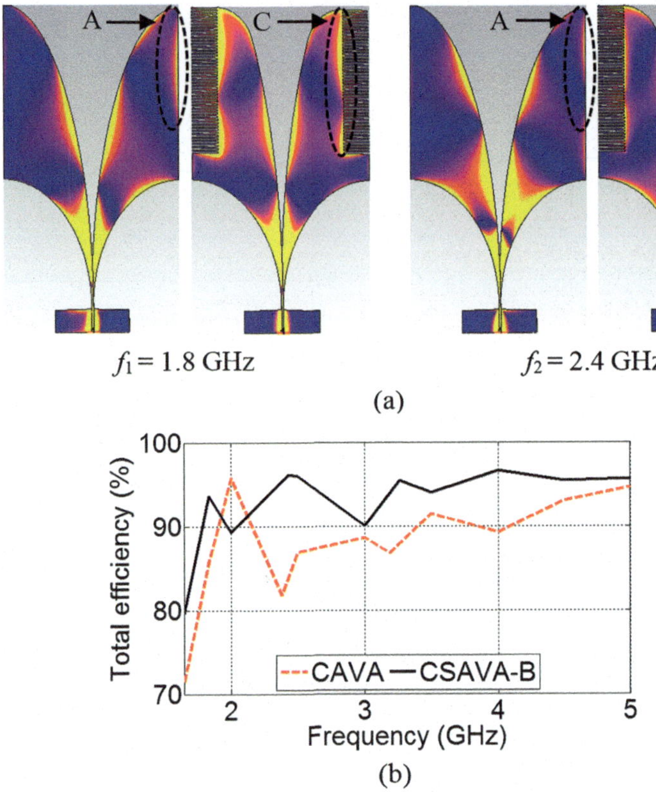

Fig. 7.3 Simulated results for the designed CAVA and CSAVA-B (**a**) surface current distribution at resonant frequencies, f_1 and f_2, indicated in Fig. 7.2b and (**b**) total radiation efficiency at lower frequencies of 1.65–5 GHz

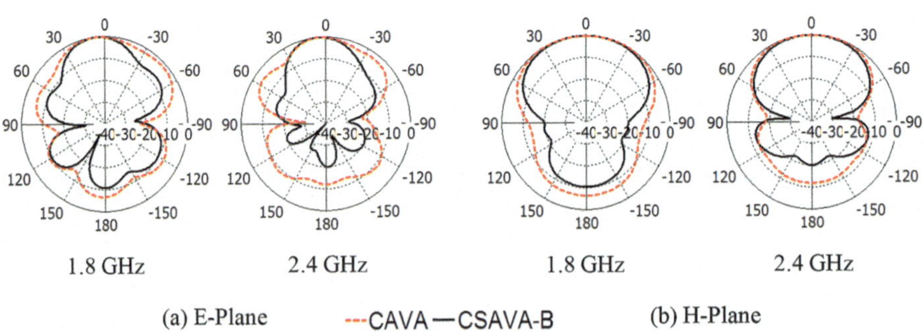

Fig. 7.4 (**a**) *E*- and (**b**) *H*-plane radiation patterns of the designed CAVA and CSAVA-B at resonant frequencies, f_1 and f_2, indicated in Fig. 7.1b

7.3 Performance of the Designed Antennas in Free Space

The gain variation versus frequency for the designed CAVA, AVA-B and CSAVA-B at the operating frequency band 1.65–18 GHz is shown in Fig. 7.5a. A zoom-in section of Fig. 7.5a at low frequencies (i.e. 1.65–5 GHz) is selected to demonstrate the

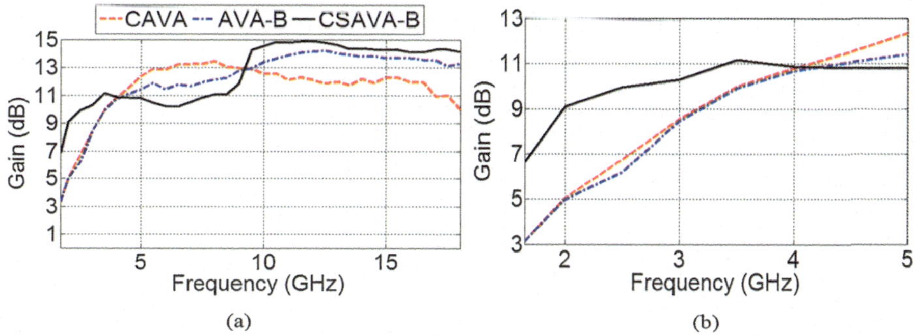

Fig. 7.5 Gain toward *Y*-direction at a bandwidth of 1.65–18 GHz and zoom-in section selected at low frequencies of 1.65–5 GHz of the designed antennas

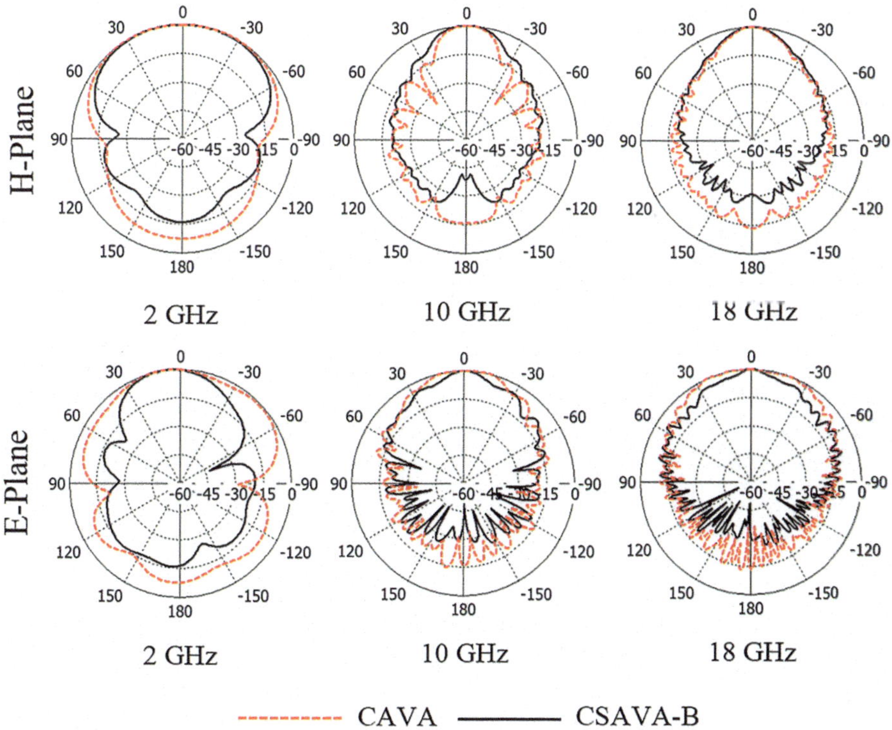

Fig. 7.6 Normalised *H*- and *E*-plane radiation patterns of the designed CAVA and CSAVA-B

Table 7.1 Gains at the lowest and highest frequencies of operating bandwidth of the referenced Vivaldi antennas and the proposed CSAVA-B

References	Dimensions (mm^2)	Bandwidth (GHz)	Gain (dB) at Lowest freq. of bandwidth	Gain (dB) at Highest freq. of bandwidth
Oktafiani et al. (2015)	67 × 70	2–20	1.8	2.8
Chu et al. (2015)	100 × 96	2.3–15	5.5	5
Siddiqui et al. (2011)	140 × 161.25	0.4–9.8	2	7.5
Li et al. (2017)	70 × 159.5	4.7–20	3.48	14.2
Wang et al. (2013)	80 × 150	2.5–15	6.9	10.8
Bayat and Mirzakhani (2012)	44 × 84	1–20	−8	2.5
Yang et al. (2010)	46 × 45	3.1–10.6	1.5	5
Hood et al. (2008)	40.16 × 42.56	3.1–10.6	−1	5
Ren et al. (2011)	46 × 69	2.25–12	5	9.5
Che et al. (2010)	43 × 25	2.5–16	0.5	6.5
Ba et al. (2014)	95 × 100	3–11	4.5	11.5
Ying and Dou (2013)	76 × 110	3.3–18	7	6
Hu et al. (2009)	126.2 × 153	2.65–12.90	9.25	11
Dastranj (2015)	31.2 × 45	6–18	8.5	10
Hong et al. (2015)	75 × 75	2–10.6	6	8.5
Huang and Hsu (2011)	78.7 × 91	3–18	5.5	10.2
Kota and Shafai (2011)	56 × 106	6–20	9	11
De Oliveira et al. (2015)	36.3 × 59.8	4.73–11	5.5	8
Natarajan et al. (2016)	60 × 60	2.4–14	4.8	10
Nassar and Weller (2015)	66 × 140	2–32	−4	0
Moosazadeh and Kharkovsky (2015)	96 × 100	2–27	2.8	11.2
Proposed antenna, CSAVA-B	120 × 202	1.65–18	6.7	14.2

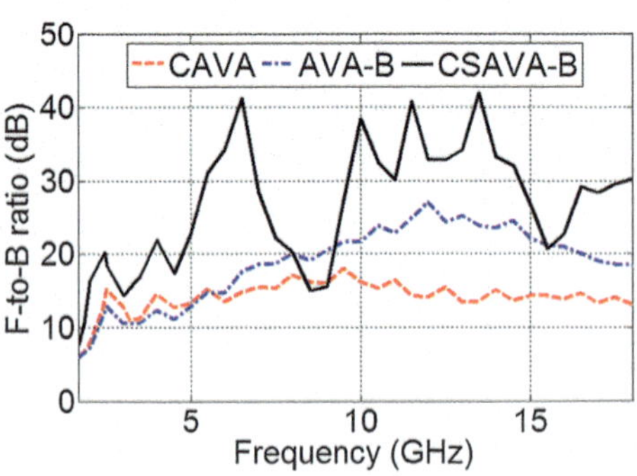

Fig. 7.7 Variation of F-to-B ratio versus frequency for the designed antennas at a bandwidth from 1.65 to 18 GHz

antenna gain improvement by the rectangular slits in periodic (comb-shaped) configuration. It can be clearly seen from Fig. 7.5b that the gain at low frequencies is higher for the designed CSAVA-B; for instance gain at 1.65 GHz is only 3 dB for the designed CAVA and AVA-B while it is 6.7 dB for the CSAVA-B.

Table 7.1 shows dimension, bandwidth, dielectric constant and gain of the referenced Vivaldi antennas and the proposed CSAVA-B. As shown in Table 7.1, the proposed CSAVA-B compared with the referenced Vivaldi antennas has the highest gain at lower frequencies, e.g. 6.7 dB at 1.65 GHz, 9.1 dB at 2 GHz and 10.3 dB at 3 GHz.

Figure 7.6 demonstrates normalised *E*-plane (*XY*-plane) and *H*-plane (*YZ*-plane) radiation patterns of the CAVA and CSAVA-B at three frequencies representing the low, central and high frequencies. It can be clearly seen from Fig. 7.6 that the designed antennas have endfire radiation patterns with the main lobe in the axial direction (i.e. *Y*-direction in Fig. 7.1c). As shown in Fig. 7.6, an improvement in F-to-B ratio for the CSAVA-B is achieved as compared to the CAVA.

The F-to-B ratio for all designed antennas is investigated as well. The F-to-B ratio versus frequency for the designed CAVA, AVA-B and CSAVA-B is shown in Fig. 7.7. As shown in Fig. 7.7, a noticeable improvement of the F-to-B ratio for the CSAVA-B is obtained compared to the CAVA and AVA-B. The maximum F-to-B ratio (42 dB) is achieved with the CSAVA-B at 13.5 GHz which is approximately 30 dB more than that for the CAVA (i.e. 13.35 dB). It should be noted that CSAVA-B has a higher F-to-B ratio at 6 GHz than the other two antennas (c.f. Fig. 7.7); however, it shows the lower gain at the same frequency (c.f. Fig. 7.5a). This conflicting phenomenon can be caused by losses imposed by slits which reduced the gain of the proposed antenna at some frequencies but increased the F-to-B ratio.

To evaluate pulse transmission and estimate signal cross-correlation, the time domain response of the proposed CSAVA-B excited with a Gaussian pulse is investigated and the results are shown in Fig. 7.8. To do this, two profiles of the CSAVA-B are put in endfire direction (*Y*-direction in Fig. 7.1c) at different far-field regions, $d = 60$ cm and 100 cm, to act as transmitting (*TX*) antenna and receiving (*RX*) antenna as shown in Fig. 7.8. The cross-correlation between transmitted signal, *TX*, and received signal, *RX*, is calculated using the fidelity factor (Lamensdorf and Susman 1994; Mehdipour et al. 2007) as follows:

$$F = \max_{\tau} \left| \frac{\int_{-\infty}^{+\infty} s(t)\, r(t-\tau)\, dt}{\sqrt{\int_{-\infty}^{+\infty} s(t)^2\, dt . \int_{-\infty}^{+\infty} r(t)^2\, dt}} \right| \tag{7.1}$$

where *s*(*t*) and *r*(*t*) are the *TX* and *RX* signals, respectively. In a research (Lamensdorf and Susman 1994), high values of F affirm a good cross-correlation between the *RX* and *TX* signals. As shown in Fig. 7.8, a similarity exists between the *RX* and the *TX* signals. The fidelity factor calculated for $d = 60$ cm is 0.85 while it is 0.9 for $d = 100$ cm. The obtained fidelity factor results reveal that the proposed CSAVA-B is able to transmit and receive signals without serious distortion.

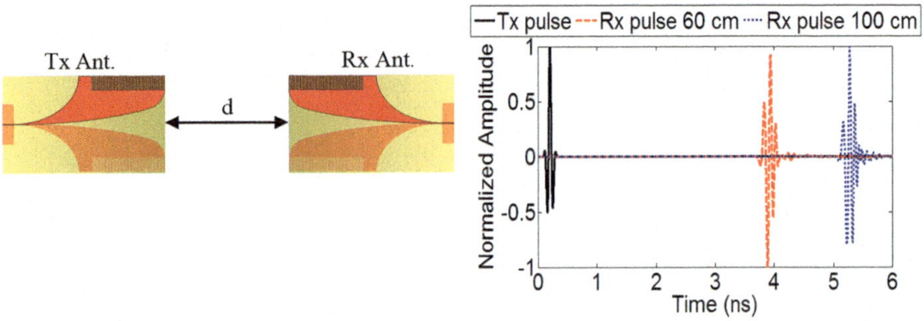

Fig. 7.8 Simulated transmitted and received pulse at the different far-field regions of the designed CSAVA-B (right) and two-antenna arrangement (left) in endfire direction (*Y*-direction shown in Fig. 7.1c)

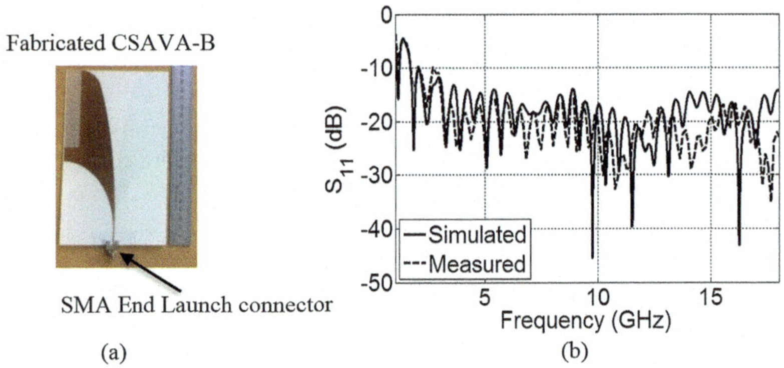

Fig. 7.9 (**a**) Photo of the fabricated CSAVA-B (top view) and (**b**) simulated and measured S_{11} of the CSAVA-B at a frequency range of 1–18 GHz (Moosazadeh et al. 2017)

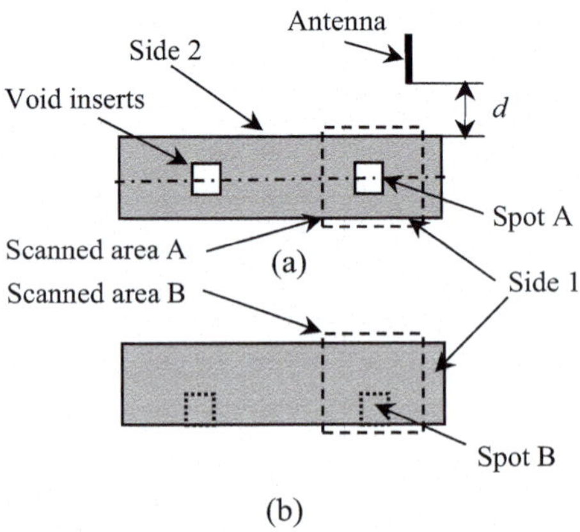

Fig. 7.10 Schematic of the concrete beam with two void inserts imaged by the proposed antenna at a standoff distance *d*: (**a**) top and (**b**) side views (Moosazadeh and Kharkovsky 2016)

7.4 Application of the CSAVA-B for Detection of Void Inside Concrete Beam

7.4.1 Fabricated Antenna and Measurement Imaging Approach

Figure 7.9a shows a prototype of the fabricated CSAVA-B fed with SMA End Launch connector. A PNA was used to measure the magnitude of the reflection coefficient of the proposed antenna. The measured magnitude of the reflection coefficient (S_{11}) in dB of the fabricated CSAVA-B is depicted in Fig. 7.9b to validate the simulated S_{11} of the antenna.

It is observed from Fig. 7.9b that the results are in good agreement at the bandwidth from 1.65 to 18 GHz for S_{11} <−10 dB except for some discrepancies at middle and higher frequencies that may be due to inaccuracy in fabrication of the antenna or arrangement of the connector.

7.4.2 Specimen Description, Simulation and Measurement Approaches

To demonstrate the applicability of the designed CSAVA-B for non-destructive testing of construction materials used in civil engineering several concrete beams possessing void mimicking foam inserts were designed, fabricated and imaged. Each concrete beam had dimensions of 400 × 100 × 100 mm^3 and two void-mimicking foam inserts. Each insert was made of a construction foam sheet and had dimensions of ~25 × 32 × 20 mm^3, which is hereafter referred to as a void insert. The antenna was attached to the scanning mechanism with a lift-off of *d* (referred to as standoff distance) as shown in Fig. 7.10a, b. Firstly, model-based numerical study of the proposed antennas and specimen under investigation was performed using a certain arrangement of the antenna and the specimen (referred to as spot measurement). The numerical study calculated the complex reflection coefficient over an entire frequency range and the electric field distribution at selected frequencies.

7.4.3 Simulation of Spot Measurement

In order to investigate the behaviour of the magnitude and phase of the reflection coefficient and the electric field intensity distribution, models of the proposed antenna and concrete beam (dielectric constant of 4.5 and loss tangent of 0.0125) with lossless void inserts (dielectric constant of 1) were created and investigated in CST. One model consists of the proposed antenna irradiating the top surface of beam as shown in Fig. 7.11a. The case when there is no void insert was also

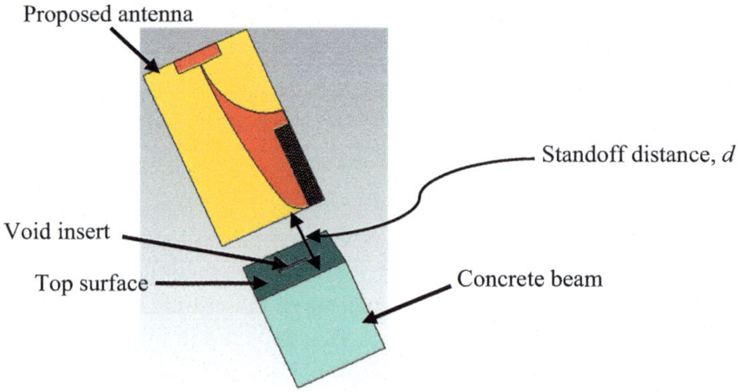

Fig. 7.11 The CST model of the proposed antenna placed at the standoff distance of *d* from above the centre of the concrete beam possessing void insert at the top surface

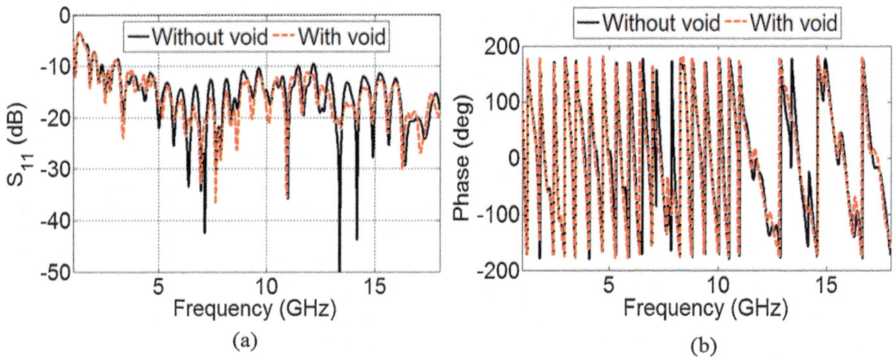

Fig. 7.12 Simulated (**a**) magnitude and (**b**) phase of the reflection coefficient with and without void inserted at the top surface in the concrete beam at a standoff distance of 10 mm

simulated and used as a reference. Two cases for standoff distance of 10 and 50 mm have been investigated in this section.

Figure 7.12a, b shows the magnitude and phase of the reflection coefficient at a frequency range of 1–18 GHz when the antenna was located at a standoff distance of 10 mm from the centre of the void insert irradiating Spot A (c.f. Fig. 7.10a). It can be seen from Fig 7.12 that there is a difference between values of the magnitude and phase of the reflection coefficient for the specimen with and without void insert over the entire frequency range from 1 to 18 GHz.

Electrical field intensity distribution for the proposed antenna and concrete beam with and without void insert at frequencies of 7.171 and 13.38 GHz is shown in Fig. 7.13. It can be seen from Fig. 7.13 that there is a significant disturbance of the electric field distribution at the interface between free space (air) and surface of concrete beam due to a high relative (to free space) dielectric constant of concrete. Then, electromagnetic waves propagate in concrete beam with multiple reflections

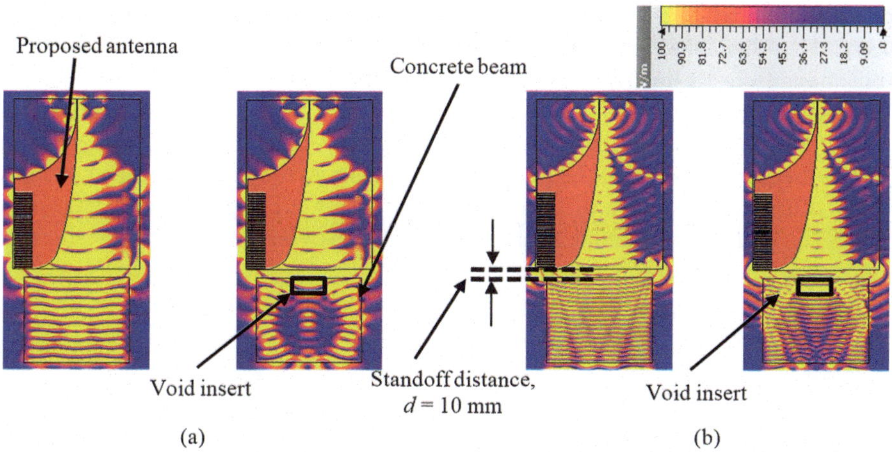

Fig. 7.13 Electrical field intensity distribution in the proposed antenna and concrete beam without and with void insert at a standoff distance of 10 mm at frequencies of (**a**) 7.171 GHz and (**b**) 13.38 GHz

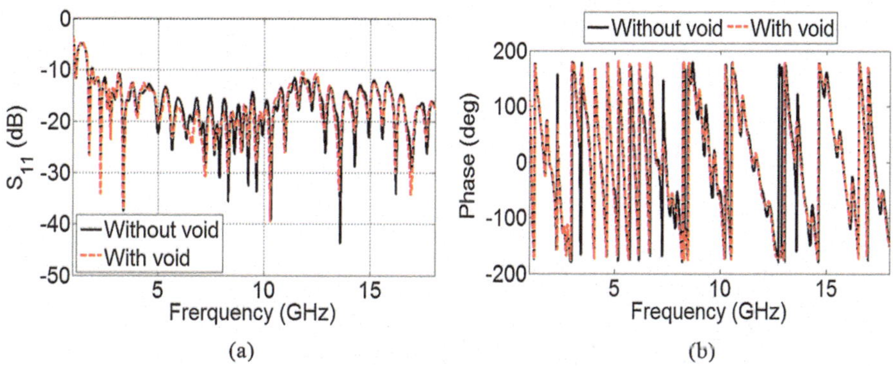

Fig. 7.14 Simulated (**a**) magnitude and (**b**) phase of the reflection coefficient with and without void inserted at the top surface in the concrete beam at a standoff distance of 50 mm

from its boundaries. Fig. 7.13 clearly shows significant changes of the electric field distribution at both frequencies when the void insert is present inside concrete.

Figure 7.14a, b shows the magnitude and phase of the reflection coefficient at a standoff distance of 50 mm at a frequency range of 1–18 GHz when the antenna was located above the concrete beam. It can be seen from Fig 7.14 that there are differences between values of the magnitude and phase of the reflection coefficient for the concrete with and without void at a frequency range from 1 to 18 GHz.

Electrical field intensity distribution for the proposed antenna and concrete beam with and without void insert at frequencies of 2.3 and 13.6 GHz is shown in Fig. 7.15. It can be seen from Fig. 7.15 that there is significant disturbance of the electric field at the interface between free space (air) and surface of concrete beam due to a high relative (to free space) dielectric constant of concrete, especially at 13.6 GHz. Then, electromagnetic waves propagate in concrete beam with multiple reflections from its boundaries and splitting of the main lobe along their propagation. Figure 7.15 clearly shows significant changes of the electric field distribution at both frequencies when the void insert is present inside concrete.

Another model included the proposed antenna irradiating Side 1 of the concrete beam at Spot B, i.e. above the side of the void insert (c.f. Fig. 7.10b) as shown in Fig. 7.16. Two cases for standoff distances of 10 and 50 mm have been investigated in this section.

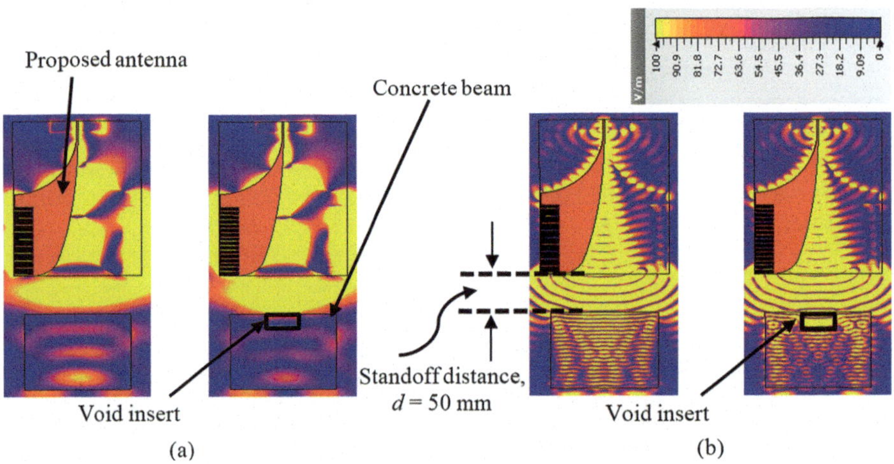

Fig. 7.15 Electrical field intensity distribution in the proposed antenna and concrete beam without and with void insert at a standoff distance of 50 mm when the antenna was located above the concrete beam at frequencies of (**a**) 2.326 GHz and (**b**) 13.6 GHz

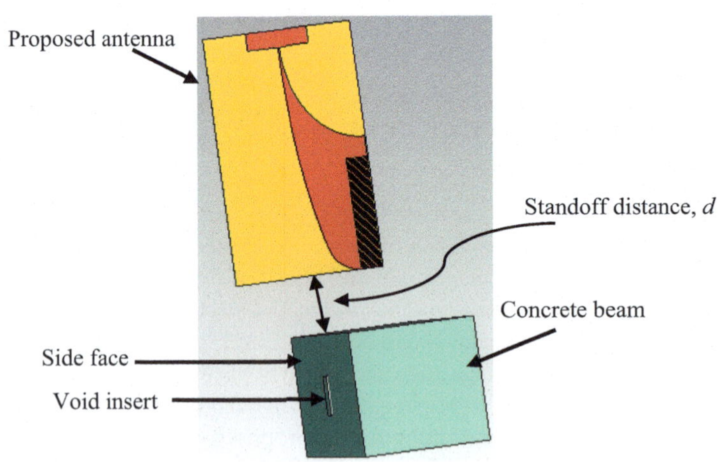

Fig. 7.16 The CST model of the proposed antenna placed at a standoff distance of d alongside the concrete beam possessing void insert

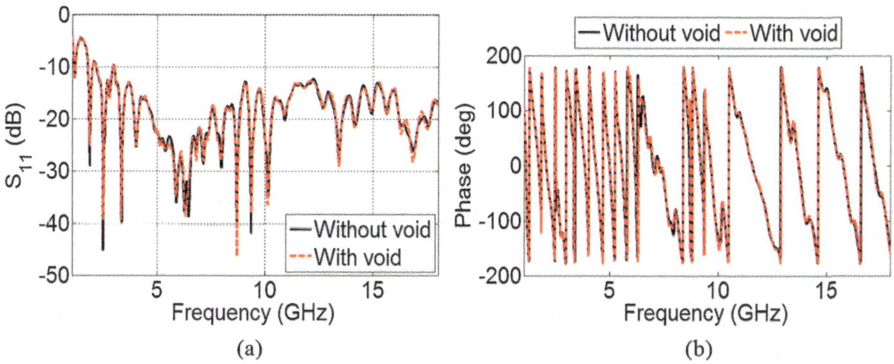

Fig. 7.17 Simulated (**a**) magnitude and (**b**) phase of the reflection coefficient with and without void inserted in the concrete beam at a standoff distance of 10 mm

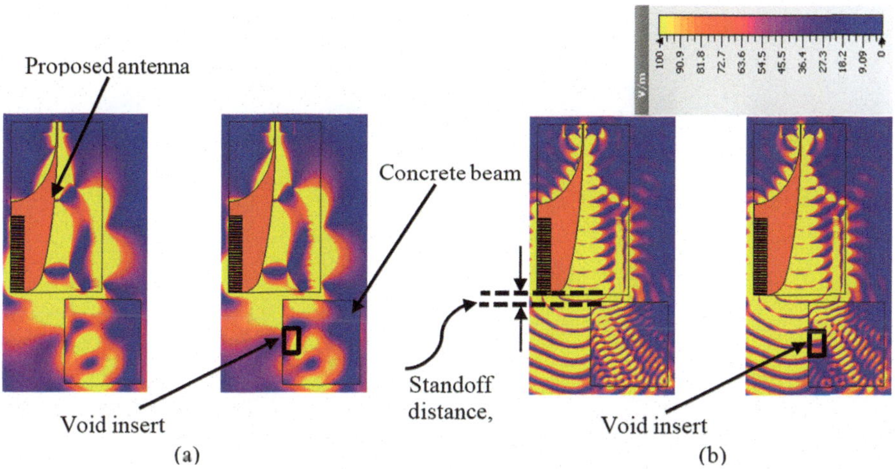

Fig. 7.18 Electrical field intensity distribution with and without void insert at a standoff distance of 10 mm when the antenna was located alongside the concrete beam at frequencies of (**a**) 1.83 GHz and (**b**) 8.718 GHz

Figure 7.17a, b shows the magnitude and phase of the reflection coefficient at a standoff distance of 10 mm at a frequency range of 1–18 GHz. It can be noticed that there is a visible difference between values of the magnitude of the reflection coefficient for the concrete with and without void insert at a frequency range from 1 to 10 GHz. As expected, this difference is smaller than the difference shown in Fig. 7.12a since the microwave signal penetrated through concrete layer before sensing the insert.

Electrical field intensity distribution for the proposed antenna and concrete beam with and without void insert at frequencies of 1.83 and 8.718 GHz is shown in Fig. 7.18. A few observations can be made from Fig. 7.18 that are similar to those

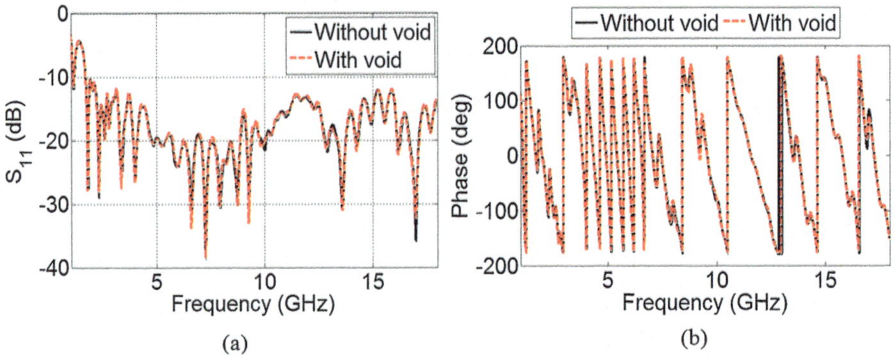

Fig. 7.19 Simulated (**a**) magnitude and (**b**) phase of the reflection coefficient with and without void inserted in the concrete beam at a standoff distance of 50 mm

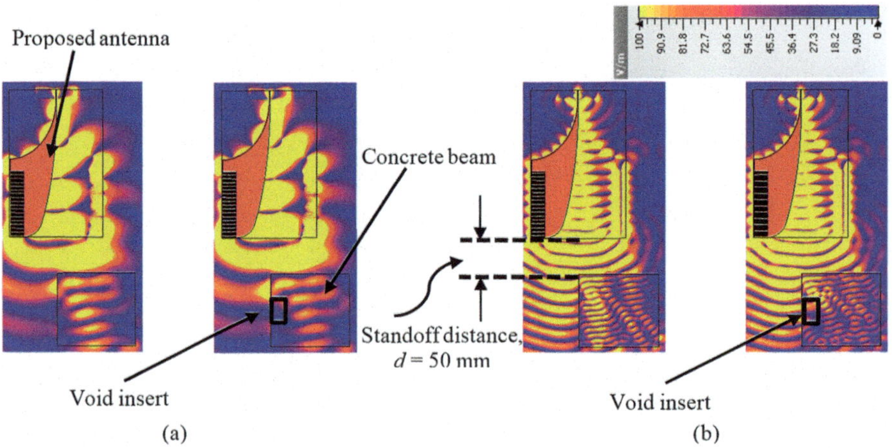

Fig. 7.20 Electrical field intensity distribution with and without void insert at a standoff distance of 50 mm when the antenna is placed alongside the concrete beam at frequencies of (**a**) 3.363 GHz and (**b**) 9.278 GHz

made for Fig. 7.13 including a significant disturbance of the electric field at the interface between free space (air) and surface of concrete and multiple reflections inside the concrete beam. Figure 7.18 also shows changes of the electric field distribution when the void insert is present inside the concrete beam. On the other hand, there are negligible changes of the magnitude of the reflection coefficient caused by the presence of the void insert at higher frequencies (>10 GHz). This can be attributed to the higher attenuation or scattering of electromagnetic waves in concrete at these frequencies.

Figure 7.19a, b shows the magnitude and phase of the reflection coefficient at a standoff distance of 50 mm at a frequency range of 1–18 GHz. Some differences can be seen between values of the magnitude and phase of the reflection coefficient for the concrete with and without void insert at middle frequencies.

Electrical field intensity distribution for the proposed antenna and concrete beam with and without void insert at frequencies of 3.363 and 9.278 GHz is shown in Fig. 7.20. A few observations can be made from Fig. 7.20 that are similar to those made from Fig. 7.18 including a significant disturbance of the electric field at the interface between free space (air) and surface of concrete and multiple reflections inside the concrete beam. Figure 7.20 shows change of the electric field distribution when void is present inside the concrete beam.

Overall, the obtained simulation results show that the proposed antenna (i.e. CSAVA-B) is able to detect void inside the concrete beam.

7.4.4 Experimental Microwave Imaging

In this section, the concrete beam was imaged at a standoff distance d using a scanned area of 150 mm by 150 mm and steps of 3 mm by 3 mm. Figure 7.21a shows a picture of one of the concrete beams tested by the proposed CSAVA-B.

First, the top surface of concrete beam with an exposed foam insert was imaged for scan area A (c.f. Fig. 7.10a). Figure 7.21b, c shows holographic slices of the 3D image obtained at standoff distances of 10 mm and 50 mm when the microwave signal was focused at the foam insert. Figure 7.21b shows the holographic slices corresponding to the top surface of foam insert, 3 and 6 mm below that surface. It can be seen from Fig. 7.21b that indications of the foam insert are visible on all three slices. On the other hand, only indications of the top of foam insert are visible at a standoff distance of 50 mm as shown in Fig. 7.21c. This can be attributed to the contribution of a strong reflection of the signal from the edges of concrete and relatively (to a standoff distance of 10 mm case) high free space losses in addition to losses in concrete.

Second, to demonstrate the ability of the imaging system with the proposed antenna for the detection of void inside concrete, Side 1 of the concrete beam was scanned for scan area B (c.f. Fig. 7.10b) and the images were generated. Holographic slices of the 3D images obtained at standoff distances of 10 mm, 50 mm and 100 mm are shown in Figs. 7.22a–c, respectively. For each case, the microwave signal was focused at the foam insert. It can be seen that the indication of the foam insert is visible at each selected standoff distance. It should also be noted that the indication of the foam insert obtained at a standoff distance of 50 mm (Fig. 7.22b) is stronger than that obtained at 10 mm (Fig. 7.22a) and 100 mm (Fig. 7.22c). The results show that the standoff distance of 50 mm can be considered as the optimal standoff distance in this case. The results also show that the penetration depth of ~50 mm in concrete has been achieved at a wide range of standoff distances in this investigation.

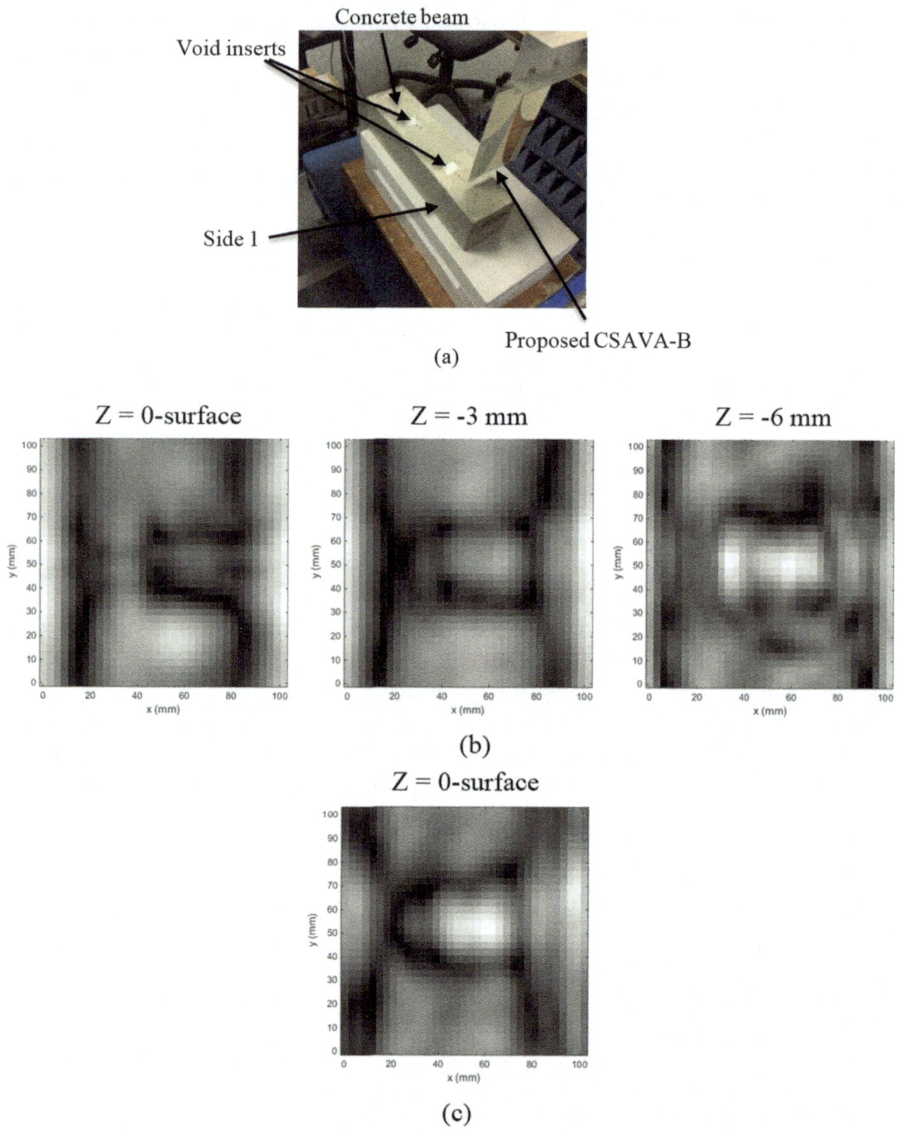

Fig. 7.21 (**a**) Picture of the proposed antenna testing the concrete beam possessing void inserts and holographic slices of 3D image obtained from the top surface of the concrete beam when microwave signal was focused at the three distances from the top surface of the beam; (**b**) 10 mm and (**c**) 50 mm

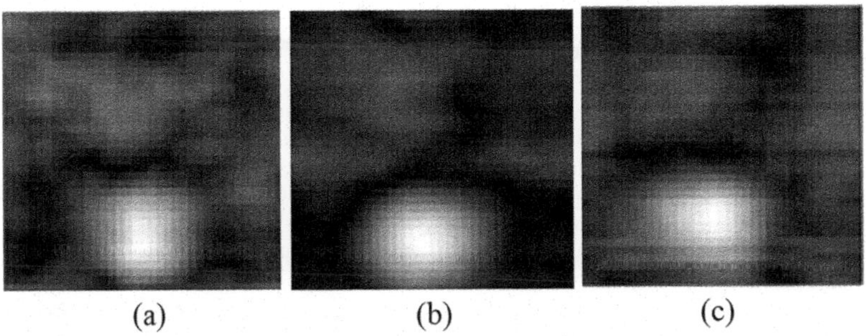

Fig. 7.22 Holographic slices of 3D image obtained from the side of the concrete beam when microwave signal was focused at the foam insert at standoff distances of (**a**) 10 mm, (**b**) 50 mm and (**c**) 100 mm (Moosazadeh et al. 2017)

7.5 Summary

An ultra-wideband elliptically tapered antipodal Vivaldi antenna with a bandwidth from 1.65 to 18 GHz was proposed, built and applied for detection of flaws such as voids in high-loss materials such as concrete. Comb-shaped slits and bending inner edges of the top and bottom radiators were applied to a conventional antipodal Vivaldi antenna to extend the operating frequency band to lower frequencies and enhance gain at low frequencies. The results show that the proposed antenna has maximum gain at low frequencies (6.7 dB at 1.65 GHz and 9.1 dB at 2 GHz) and front-to-back ratio of 42 dB at 13.5 GHz. Applicability of the proposed antenna along with an imaging system for the purpose of the detection of voids inside a concrete beam is demonstrated.

References

Ba, H. C., Shirai, H., & Ngoc, C. D. (2014). *Analysis and design of antipodal Vivaldi antenna for UWB applications*. In 2014 IEEE Fifth International Conference on Communications and Electronics (ICCE). IEEE, pp. 391–394.

Bayat, A., & Mirzakhani, R. (2012). A parametric study and design of the Balanced Antipodal Vivaldi Antenna (BAVA). *Session 2P8 Mobile Antennas, Printed Antennas, and Array Antennas*, 452.

Che, Y., Li, K., Hou, X., & Tian W. (2010). *Simulation of a small sized antipodal Vivaldi antenna for UWB applications*. In 2010 IEEE International Conference on Ultra-Wideband (ICUWB). IEEE, pp. 1–3.

Chu, H. B., Shirai, H., & Dao, C. N. (2015). *Effect of curvature of antipodal structure on Vivaldi antennas*. In 2015 IEEE International Symposium on Antennas and Propagation & USNC/URSI National Radio Science Meeting. IEEE, pp. 2331–2332.

Dastranj, A. (2015). Wideband antipodal Vivaldi antenna with enhanced radiation parameters. *IET Microwaves, Antennas and Propagation, 9*, 1755–1760.

De Oliveira, A. M., Perotoni, M. B., Kofuji, S. T., & Justo, J. F. (2015). A palm tree antipodal Vivaldi antenna with exponential slot edge for improved radiation pattern. *Antennas and Wireless Propagation Letters, IEEE, 14*, 1334–1337.

Hong, H., Ahn, J., Jeong, J.-G., & Yoon, Y. J. (2015). *Gain enhancement technique for an antipodal Vivaldi antenna*. In 2015 IEEE International Symposium on Antennas and Propagation & USNC/URSI National Radio Science Meeting. IEEE, pp. 2343–2344.

Hood, A. Z., Karacolak, T., & Topsakal, E. (2008). A small antipodal Vivaldi antenna for ultrawideband applications. *Antennas and Wireless Propagation Letters, IEEE, 7*, 656–660.

Hu, S., Dou, W., & Law, C. (2009). A tapered slot antenna with flat and high gain for ultrawideband applications. *Journal of Electromagnetic Waves and Applications, 23*, 723–728.

Huang, T.-J., & Hsu, H.-T. (2011). *Antipodal dual exponentially tapered slot antennas (DETSA) with corrugations for front-to-back ratio improvement*. In 2011 IEEE International Workshop on Electromagnetics, Applications and Student Innovation (iWEM). IEEE, pp. 48–51

Kota, K., & Shafai, L. (2011). Gain and radiation pattern enhancement of balanced antipodal Vivaldi antenna. *Electronics Letters, 47*, 1.

Lamensdorf, D., & Susman, L. (1994). Baseband-pulse-antenna techniques. *IEEE Antennas and Propagation Magazine, 36*, 20–30.

Li, E., Wang, C., & Guo, G. (2017). Radiation enhanced Vivaldi antenna with double-antipodal structure. *IEEE Antennas and Wireless Propagation Letters, 16*, 561–564.

Mehdipour, A., Mohammadpour-Aghdam, K., & Faraji-Dana, R. (2007). Complete dispersion analysis of Vivaldi antenna for ultra wideband applications. *Progress in Electromagnetics Research, 77*, 85–96.

Moosazadeh, M., & Kharkovsky, S. (2015, October 4–7). *Design of ultra-wideband antipodal Vivaldi antenna for microwave imaging applications*. In 2015 IEEE International Conference on Ubiquitous Wireless Broadband (ICUWB), pp. 1–4.

Moosazadeh, M., Kharkovsky, S., Case, J. T., & Samali, B. (2017). Antipodal Vivaldi antenna with improved radiation characteristics for civil engineering applications. *IET* Microwaves, *Antennas and Propagation, 11*(6), 796–803.

Nassar, I. T., & Weller, T. M. (2015). A novel method for improving antipodal Vivaldi antenna performance. *IEEE Transactions on Antennas and Propagation, 63*, 3321–3324.

Natarajan, R., George, J. V., Kanagasabai, M., Lawrance, L., Moorthy, B., Rajendran, D. B., & Alsath, M. G. N. (2016). Modified antipodal Vivaldi antenna for ultra-wideband communications. *IET Microwaves, Antennas and Propagation, 10*(4), 401–405.

Oktafiani, F., Amrullah, Y., Saputera, Y., Wahyu, Y., & Wijayanto, Y. (2015). *Analysis of corrugated edge variations on balanced antipodal Vivaldi antennas*. In 2015 International Conference on Radar, Antenna, Microwave, Electronics, and Telecommunications (ICRAMET). IEEE, pp. 1–5.

Ren, F.-C., Zhang, F.-S., Chen, B., & Zhou, Q.-C. (2011). *Compact tapered slot antenna for wideband applications*. In 2011 IEEE CIE International Conference on Radar (Radar). IEEE, pp. 1161–1163.

Siddiqui, J., Antar, Y., Freundorfer, A., Smith, E., Morin, G., & Thayaparan, T. (2011). Design of an ultrawideband antipodal tapered slot antenna using elliptical strip conductors. *IEEE Antennas and Wireless Propagation Letters, 10*, 251–254.

Wang, Y.-W., Wang, G.-M., & Zong, B.-F. (2013). Directivity improvement of Vivaldi antenna using double-slot structure. *IEEE Antennas and Wireless Propagation Letters, 12*, 1380–1383.

Yang, L., Guo, H., Liu, X., Du, H., & Ji, G. (2010). *An antipodal Vivaldi antenna for ultra-wideband system*. In 2010 IEEE International Conference on Ultra-Wideband (ICUWB). IEEE, pp. 1–4.

Ying, Q., & Dou, W. (2013). *Simulation of two compact antipodal Vivaldi antennas with Radiation Characteristics enhancement*. In 2013 Proceedings of the International Symposium on Antennas & Propagation (ISAP). IEEE, pp. 523–526.

Chapter 8
Conclusion and Suggestions for Future Work

8.1 Conclusion

In this book, antipodal Vivaldi antennas for microwave and/or millimetre wave imaging of construction materials and structures have been developed. In the developed AVAs, different techniques including slits and dielectric lenses were utilised to achieve desired radiation characteristics including extended low end of frequency band, high gain and F-to-B ratio, E- and H-plane radiation patterns, E-plane tilt of beam and E-plane HPBW. The results of parametric studies and design of these AVAs were presented. The proposed AVAs were fabricated and tested. The fabricated AVAs were applied for imaging of different construction materials with respect to their achieved operating frequency ranges. This chapter presents the main outcomes of this research and suggestions for future work.

A small and modified AVA with small size of 30 × 55 × 0.508 mm^3 (i.e. $0.5\,\lambda \times 0.9\,\lambda \times 0.005\,\lambda$, λ is the free-space wavelength corresponding to the lowest operating frequency of 5 GHz) with a bandwidth from 5 to 50 GHz was developed. Lower end of frequency band limitation for $S_{11} < -10$ dB was achieved by rectangular slits at sun-shaped configuration. The applied sun-shaped elements enhance antenna gain up to 5.7 dB at lower frequencies. A half-elliptical shaped dielectric lens as extension of antenna substrate is employed to enhance antenna gain at higher frequencies, and as a result the antenna gain is 14 dB at 50 GHz. In addition, the proposed antenna demonstrates reduced E-plane tilt of beam, e.g. −1° at 50 GHz, low side lobe level, cross-polarisation level (<−22 dB), E-plane half-power beamwidth (e.g. 12.2° at 50 GHz) and improved F-to-B ratio. The antenna has a gain >5.7 dB at the frequency range of 5–50 GHz, >10 dB for the range of 10–50 GHz and >13 dB for the millimetre wave frequency range of 30–50 GHz. The imaging results show that the proposed antenna can be used for high-range resolution imaging of a variety of low-loss construction materials as well as other materials.

An AVA with high gain and F-to-B ratio, and wide operating bandwidth from 3.4 to 40 GHz, was designed. It is based on periodic slit-edge technique to lower cut-off

M. Moosazadeh, *Antipodal Vivaldi Antennas for Microwave Imaging of Construction Materials and Structures*,
https://doi.org/10.1007/978-3-030-05566-0_8

frequency of CAVA and enhance CAVA gain at low frequencies. To increase antenna gain at higher frequencies, a trapezoid-shaped dielectric lens as an extension of substrate is employed. The narrow E-plane HPBW and small E-plane tilt of beam up to 40 GHz are the main features of the proposed antenna. The specification and performances of the referenced Vivaldi antennas in terms of size, operation bandwidth, relative permittivity, and gain and F-to-B ratio in comparison with the proposed antenna were demonstrated. Applicability of the proposed antenna for UWB imaging of construction materials and structures is demonstrated using three different samples including a concrete block with plasterboard sheets to highlight the high-range resolution of the antenna.

A modified AVA was proposed and designed. To extend the low end of frequency band of the CAVA, the inner edges of the top and bottom radiators of the CAVA have been bent; however, the AVA-B gain was low at lower frequencies. To enhance gain at lower frequencies, regular triangular slit was applied to the AVA-B. Finally, applicability of the proposed antenna for UWB imaging of construction materials and structures for non-destructive testing and evaluation of concrete-based specimens was demonstrated to highlight the capability of the proposed antenna with the imaging system for the detection of flaws and structural features located in relatively small depth of concrete with high-range resolution (a few millimetres) and deep penetration depth and propagation path (more than 200 mm in two ways).

A miniaturised AVA with a minimum operating frequency of 1 GHz and high gain for the entire frequency range of 1–30 GHz was described. The CAVA was modified to design a small AVA with reduced size. To extend the low end of frequency band and further reduce the size of antenna, the inner edges of the top and bottom radiators of the designed small AVA have been bent. To enhance the gain at lower frequencies, regular slit edge was designed. Finally, a half-elliptical shaped dielectric lens as an extension of the antenna substrate was used to feature high gain and F-to-B ratio. The capability of the proposed antenna for UWB imaging was demonstrated for the detection of voids inside concrete. The result shows that range resolution of a few millimetres in concrete has been achieved at a wide range of standoff distances.

A high-gain and -F-to-B-ratio comb-shaped slit AVA was presented. Inner edges of the top and bottom radiators of the conventional antipodal Vivaldi antenna have been properly bent to extend the low end of frequency band. Rectangular slits in periodic (comb-shaped) configuration are applied to the antenna to enhance gain at low frequencies. The proposed antenna has impedance bandwidth of 1.65–18 GHz, high gain at low frequencies (6.7 dB at 1.65 GHz and 9.1 dB at 2 GHz) and F-to-B ratio of 42 dB at 13.5 GHz. The proposed antenna was employed to microwave imaging system for the purpose of detection of void inside the concrete beam.

Bandwidth, size and designed techniques of AVAs developed in the book along with recommendation for materials and defects are summarised in Table 8.1. As shown in Table 8.1 the operating frequency range of the developed antennas is limited by 1 GHz at lower end of the range. It may not be enough to achieve the desired penetration depth and range resolution in high-loss materials such as concrete.

Table 8.1 Bandwidth, size and designed techniques of AVAs developed in the book and recommendation for materials and defects

AVAs/designed techniques	Bandwidth (GHz)	Size (mm^2)	Recommendation for materials/defects
MAVA-HEDL/rectangular slits at a sun-shaped configuration and half-elliptical diclectric lens	5–50	30 × 55	Low-loss materials (plasterboard, polymer, glass)/impact damage, cracks, delamination
PSEAVA-TDL/periodic slit edge and trapezoid-shaped dielectric lens	3.4–40	40 × 90	Low-loss and high-loss materials (plasterboard layered structures, wood, polymers)/surface damages of concrete under low-loss material coating
RTSAVA-B/regular triangular slit in saw-shaped configuration and bending	2–27	96 × 100	Low-loss and high-loss materials (concrete structures/cavities, flaws at small depth)
MRSEAVA-HEDL/regular slit edge, bending and half-elliptical dielectric lens	1–30	96 × 140	Low-loss and high-loss materials (concrete, polymers, layered structures/voids, cracks)
CSAVA-B/rectangular slits in periodic (comb-shaped) configuration and bending	1.65–18	120 × 202	Low-loss and high-loss materials (concrete, layered structures/voids)

8.2 Suggestions for Future Work

Suggestions for future work are the following:

(a) Employing the proposed AVAs for imaging of different materials such as geo-polymer concrete, masonry blocks and wood
(b) Considering novel techniques for wideband gain enhancement of AVA for microwave and/or millimetre wave imaging applications
(c) Investigating real-time microwave imaging system for detection of defects inside construction materials and structures
(d) Developing UWB AVA array for microwave imaging systems
(e) Developing UWB AVAs at operating frequencies lower than 1 GHz to provide deeper penetration depth for detection of defects inside high-loss materials and structures such as reinforecd concrete

Index

M. Moosazadeh, *Antipodal Vivaldi Antennas for Microwave Imaging of Construction Materials and Structures*,
https://doi.org/10.1007/978-3-030-05566-0

Printed by Printforce, the Netherlands